Elliott Coues

A check list of North American birds

Elliott Coues

A check list of North American birds

ISBN/EAN: 9783337717148

Printed in Europe, USA, Canada, Australia, Japan

Cover: Foto ©ninafisch / pixelio.de

More available books at **www.hansebooks.com**

A CHECK LIST

OF

NORTH AMERICAN BIRDS.

By

ELLIOTT COUES.

SALEM.
NATURALISTS' AGENCY.
1873.

PRINTED AT
THE SALEM PRESS,
F. W. PUTNAM & CO.,
Proprietors.

CHECK LIST OF NORTH AMERICAN BIRDS.

NOTE. — The species are numbered consecutively from 1 to 635. Stragglers have the number in brackets. Varieties bear the number of the species to which they belong, with *a*, *b*, *c*, etc., unless a variety is our only representative of the species, when it is separately enumerated. Obscure or doubtful species are marked with a note of interrogation after the number. Each species is followed by the original describer's name; when this is not also the authority for the nomenclature adopted the name of such authority is added, the former being retained in parenthesis. A similar practice is observed in the cases of varieties; when, as in most instances, they were originally described as species they are followed by the authority for their reduction to varieties, as well as by the name of the describer; the latter in parenthesis.

The List contains a very few species discovered since the "Key" was printed; otherwise, it is an exact reflection of that work, the arrangement and nomenclature being identical. The numbers of the genera as used in the Key are given in the head lines of the List in order to facilitate reference.

Authors are at variance in the formation of the genitive of Latinized proper names; in the absence of any universally observed rule, euphony may perhaps be advantageously consulted. In the Key, the *i* was doubled in all cases of words ending in a consonant, the nominative being considered to end in *-ius;* this practice is preferably applicable to monosyllables, as *Bairdii*, and polysyllables, as *Audubonii*. But it is necessary to use single *i* in words ending in *r*, as *Cooperi*, and best to do so in most cases of dissyllables, as *Wilsoni*, *Cassini*, *Swainsoni*. The same is the case with all words ending in a vowel.

The following are the abbreviations used for authors' names occurring most frequently; others are for the most part written in full:—
All., Allen; *Aud.*, Audubon; *Bd.*, Baird; *Bodd.*, Boddaert; *Bp.*, Bonaparte; *Cab.*, Cabanis; *Cass.*, Cassin; *Cs.*, Coues; *Gamb.*, Gambel; *Gm.*, Gmelin; *Gr.*, Gray; *L.*, Linnæus; *Lafr.*, Lafresnaye; *Lath.*, Latham; *Lawr.*, Lawrence; *Licht.*, Lichtenstein; *Nutt.*, Nuttall; *Reich.*, Reichenbach; *Ridg.*, Ridgway; *Scl.*, Sclater; *Steph.*, Stephens; *Sw.*, Swainson; *Temm.*, Temminck; *Towns.*, Townsend; *V.*, Vieillot; *Vig.*, Vigors; *Wagl.*, Wagler; *Wils.*, Wilson.

GEN. 1–3 OF KEY.   5

1. **TURDUS MIGRATORIUS** L.
   Robin.

   1a. **TURDUS MIGRATORIUS** L.,
       var. CONFINIS (Bd.) Cs.
       St. Lucas Robin.

2. **TURDUS NÆVIUS** GM.
   Varied Thrush.

3. **TURDUS MUSTELINUS** GM.
   Wood Thrush.

4. **TURDUS PALLASI** CAB.
   Hermit Thrush.

   4a. **TURDUS PALLASI** CAB.,
       var. AUDUBONII (Bd.) Cs.
       Audubon's Thrush.

   4b. **TURDUS PALLASI** CAB.,
       var. NANUS (Aud.) Cs.
       Dwarf Thrush.

5. **TURDUS SWAINSONI** CAB.
   Olive-backed Thrush.

   5a. **TURDUS SWAINSONI** CAB.,
       var. ALICIÆ (Bd.) Cs.
       Alice's Thrush.

   5b. **TURDUS SWAINSONI** CAB.,
       var. USTULATUS (Nutt.) Cs.
       Oregon Thrush.

6. **TURDUS FUSCESCENS** STEPH.
   Wilson's Thrush, Veery.

7. **OREOSCOPTES MONTANUS** (TOWNS.) BD.
   Mountain Mockingbird.

8. **MIMUS POLYGLOTTUS** (L.) BOIE.
   Mockingbird.

9. **MIMUS CAROLINENSIS** (L.) Gr.
   Catbird.

10. **HARPORHYNCHUS RUFUS** (L.) Cab.
    Brown Thrush. Thrasher.

10a. **HARPORHYNCHUS RUFUS** (L.) Cab.,
   var. LONGIROSTRIS (Lafr.) Cs.
   Long-billed Thrush.

11. **HARPORHYNCHUS CURVIROSTRIS** (Sw.) Cab.,
    var. PALMERI Ridg.
    Curve-billed Thrush.

11bis.* **HARPORHYNCHUS BENDIREI** Cs.
   Bendire's Thrush.

12. **HARPORHYNCHUS CINEREUS** Bd.
    Cinereous Thrush.

13. **HARPORHYNCHUS REDIVIVUS** (Gamb.) Cab.
    Sickle-billed Thrush.

13a. **HARPORHYNCHUS REDIVIVUS** (Gamb.) Cab.,
   var. LECONTEI (Lawr.) Cs.
   LeConte's Thrush.

14. **HARPORHYNCHUS CRISSALIS** Henry.
    Red-vented Thrush.

[15]. **SAXICOLA ŒNANTHE** (L.) Bechstein.
    Stone Chat. Wheat-ear.

16. **SIALIA SIALIS** (L.) Haldeman.
    Eastern Bluebird.

17. **SIALIA MEXICANA** Sw.
    Western Bluebird.

18. **SIALIA ARCTICA** Sw.
    Arctic Bluebird.

---

*11bis. Not in the Key. See Am. Nat., Vol. vii, p. 330, 1873.

19. CINCLUS MEXICANUS Sw.
    Water Ouzel. Dipper.

[20]. PHYLLOPNEUSTE BOREALIS Blasius.
    Kennicott's Sylvia.

21. REGULUS CALENDULA (L.) Licht.
    Ruby-crowned Kinglet.

22. REGULUS SATRAPA Licht.
    Golden-crested Kinglet.

23. POLIOPTILA CÆRULEA (L.) Scl.
    Blue-gray Gnatcatcher.

24. POLIOPTILA MELANURA Lawr.
    Black-headed Gnatcatcher.

25. POLIOPTILA PLUMBEA Bd.
    Plumbeous Gnatcatcher.

26. CHAMÆA FASCIATA Gamb.
    Fasciated Tit. Ground Wren.

27. LOPHOPHANES BICOLOR (L.) Bp.
    Tufted Titmouse.

28. LOPHOPHANES INORNATUS (Gamb.) Cass.
    Plain Titmouse.

29. LOPHOPHANES ATRICRISTATUS Cass.
    Black-crested Titmouse.

30. LOPHOPHANES WOLLWEBERI Bp.
    Bridled Titmouse.

31. PARUS ATRICAPILLUS L.
    Black-capped Chickadee.

31a. PARUS ATRICAPILLUS L.,
    var. septentrionalis (Harris) All.
    Long-tailed Chickadee.

31b. **PARUS ATRICAPILLUS** L.,
var. CAROLINENSIS (Aud.) Cs.
Carolina Chickadee.

31c. **PARUS ATRICAPILLUS** L.,
var. OCCIDENTALIS (Bd.) Cs.
Western Chickadee.

32. **PARUS MONTANUS** GAMB.
Mountain Chickadee.

33. **PARUS HUDSONICUS** FORSTER.
Hudsonian Chickadee.

34. **PARUS RUFESCENS** TOWNS.
Chestnut-backed Chickadee.

✓ 35. **PSALTRIPARUS MINIMUS** (TOWNS.) BP.
Least Titmouse.

36. **PSALTRIPARUS PLUMBEUS** BD.
Plumbeous Titmouse.

✓ 37. **AURIPARUS FLAVICEPS** (SUND.) BD.
Yellow-headed Titmouse.

✓ 38. **SITTA CAROLINENSIS** GM.
White-bellied Nuthatch.

38a. **SITTA CAROLINENSIS** GM.,
var. ACULEATA (Cass.) All.
Slender-billed Nuthatch.

39. **SITTA CANADENSIS** L.
Red-bellied Nuthatch.

40. **SITTA PUSILLA** LATH.
Brown-headed Nuthatch.

✓ 41? **SITTA PYGMÆA** VIG.
Pygmy Nuthatch.

✓ 42. **CERTHIA FAMILIARIS** L.
Brown Creeper.

✓ 43. CAMPYLORHYNCHUS
BRUNNEICAPILLUS (Lafr.) Gr.
Brown-headed Creeper-wren.

✓ 44. CAMPYLORHYNCHUS AFFINIS Xantus.
Allied Creeper-wren.

✓ 45. SALPINCTES OBSOLETUS (Say) Cab.
Rock Wren.

46. CATHERPES MEXICANUS (Sw.) Bd.
White-throated Wren.

✓ 47. THRYOTHORUS LUDOVICIANUS (Gm.) Bp.
Great Carolina Wren.

47a. THRYOTHORUS LUDOVICIANUS (Gm.) Bp.,
var. berlandieri (Couch) Cs.
Berlandier's Wren.

48. THRYOTHORUS BEWICKII (Aud.) Bp.
Bewick's Wren.

48a. THRYOTHORUS BEWICKII (Aud.) Bp.
var. leucogaster (Gould) Bd.
White-bellied Wren.

48b. THRYOTHORUS BEWICKII (Aud.) Bp.,
var. spilurus (Vig.) Bd.
Speckled-tailed Wren.

✓ 49. TROGLODYTES AEDON V.
House Wren.

✓ 49a. TROGLODYTES AEDON V.,
var. parkmanni (Aud.) Cs.
Western House Wren.

50. ANORTHURA TROGLODYTES (L.) Cs.,
var. hyemalis (Wils.) Cs.
Winter Wren.

50a. ANORTHURA TROGLODYTES (L.) Cs.,
var. ALASCENSIS (Bd.) Cs.
Alaskan Wren.

✓ 51. TELMATODYTES PALUSTRIS (Wils.) Cab.
Long-billed Marsh Wren.

✓ 52. CISTOTHORUS STELLARIS (Licht.) Cab.
Short-billed Marsh Wren.

✓ 53. EREMOPHILA ALPESTRIS (Forst.) Boie.
Horned Lark; Shore Lark.

53a. EREMOPHILA ALPESTRIS (Forst.) Boie,
var. CHRYSOLÆMA (Wagl.) Bd.
South-western Lark.

54. BUDYTES FLAVA (L.) Cuv.
Yellow Wagtail.

✓ 55. ANTHUS LUDOVICIANUS (Gm.) Licht.
Brown Lark; Titlark; Pipit.

56. NEOCORYS SPRAGUEI (Aud.) Scl.
Missouri Skylark.

✓ 57. MNIOTILTA VARIA (L.) V.
Black-and-white Creeper.

58. PARULA AMERICANA (L.) Bp.
Blue yellow-backed Warbler.

✓ 59. PROTONOTARIA CITRÆA (Bodd.) Bd.
Prothonotary Warbler.

60. HELMITHERUS VERMIVORUS (Gm.) Bp.
Worm-eating Warbler.

61. HELMITHERUS SWAINSONI (Aud.) Bp.
Swainson's Warbler.

✓ 62. HELMINTHOPHAGA PINUS (L.) Bd.
Blue-winged Yellow Warbler.

GEN. 34–35 OF KEY.

✓ 63. **HELMINTHOPHAGA CHRYSOPTERA** (L.) BD.
       Blue Golden-winged Warbler.

  64. **HELMINTHOPHAGA BACHMANI** (AUD.) CAB.
       Bachman's Warbler.

  65. **HELMINTHOPHAGA LUCIÆ** COOP.
       Lucy's Warbler.

  66. **HELMINTHOPHAGA VIRGINIÆ** BD.
       Virginia's Warbler.

✓ 67. **HELMINTHOPHAGA RUFICAPILLA** (WILS.) BD.
       Nashville Warbler.

✓ 68. **HELMINTHOPHAGA CELATA** (SAY) BD.
       Orange-crowned Warbler.

  69. **HELMINTHOPHAGA PEREGRINA** (WILS.) CAB.
       Tennessee Warbler.

  70. **DENDRŒCA ÆSTIVA** (GM.) BD.
       Summer Warbler.

  71. **DENDRŒCA VIRENS** (GM.) BD.
       Black-throated Green Warbler.

  72. **DENDRŒCA OCCIDENTALIS** (TOWNS.) BD.
       Western Warbler.

  73. **DENDRŒCA TOWNSENDII** (NUTT.) BD.
       Townsend's Warbler.

  74. **DENDRŒCA CHRYSOPAREIA** SCL. ET SALV.
       Golden-cheeked Warbler.

  75. **DENDRŒCA NIGRESCENS** (TOWNS.) BD.
       Black-throated Gray Warbler.

  76. **DENDRŒCA CÆRULESCENS** (L.) BD.
       Black-throated Blue Warbler.

  77. **DENDRŒCA CÆRULEA** (WILS.) BD.
       Cærulean Warbler.

78. DENDRŒCA CORONATA (L.) Gr.
Yellow-rumped Warbler; Myrtle Bird.

79. DENDRŒCA AUDUBONII (Towns.) Bd.
Audubon's Warbler.

80. DENDRŒCA BLACKBURNIÆ (Gm.) Bd.
Blackburnian Warbler.

81. DENDRŒCA STRIATA (Forst.) Bd.
Black-poll Warbler.

82. DENDRŒCA CASTANEA (Wils.) Bd.
Bay-breasted Warbler.

83. DENDRŒCA PENNSYLVANICA (L.) Bd.
Chestnut-sided Warbler.

84. DENDRŒCA MACULOSA (Gm.) Bd.
Black-and-Yellow Warbler.

85. DENDRŒCA TIGRINA (Gm.) Bd.
Cape May Warbler.

86. DENDRŒCA DISCOLOR (V.) Bd.
Prairie Warbler.

87. DENDRŒCA GRACIÆ Coues.
Grace's Warbler.

88. DENDRŒCA DOMINICA (L.) Bd.
Yellow-throated Warbler.

89. DENDRŒCA KIRTLANDI Bd.
Kirtland's Warbler.

90. DENDRŒCA PALMARUM (Gm.) Bd.
Yellow Red-poll Warbler.

91. DENDRŒCA PINUS (Wils.) Bd.
Pine-creeping Warbler.

92. SEIURUS AUROCAPILLUS (L.) Sw.
Golden-crowned Thrush.

93. SEIURUS NOVEBORACENSIS (Gm.) Nutt.
Water Wagtail; Water Thrush.

94. SEIURUS LUDOVICIANUS (V.) Bp.
Large-billed Water Thrush.

95. OPORORNIS AGILIS (Wils.) Bd.
Connecticut Warbler.

96. OPORORNIS FORMOSUS (Wils.) Bd.
Kentucky Warbler.

97. GEOTHLYPIS TRICHAS (L.) Cab.
Maryland Yellow-throat.

98. GEOTHLYPIS PHILADELPHIA (Wils.) Bd.
Mourning Warbler.

99?* GEOTHLYPIS MACGILLIVRAYI (Aud.) Bd.
Macgillivray's Warbler.

100 ICTERIA VIRENS (L.) Bd.
Yellow-breasted Chat.

100a. ICTERIA VIRENS (L.) Bd.,
var. longicauda (Lawr.) Cs.
Long-tailed Chat.

101. MYIODIOCTES MITRATUS (Gm.) Aud.
Hooded Flycatching Warbler.

102. MYIODIOCTES PUSILLUS (Wils.) Bp.
Green Black-capped Flycatching Warbler.

103. MYIODIOCTES CANADENSIS (L.) Aud.
Canadian Flycatching Warbler.

104. SETOPHAGA RUTICILLA (L.) Sw.
Redstart.

105. SETOPHAGA PICTA Sw.
Painted Flycatcher.

---

* This is probably only a variety of 98.

✓106. CERTHIOLA FLAVEOLA (L.) SUND.
  Honey Creeper.

✓107. PYRANGA RUBRA (L.) V.
  Scarlet Tanager.

108. PYRANGA ÆSTIVA (L.) V.
  Summer Redbird.

108a. PYRANGA ÆSTIVA (L.) V.,
  var. COOPERI (Ridg.) Cs.
  Cooper's Tanager.

109. PYRANGA HEPATICA Sw.
  Hepatic Tanager.

110. PYRANGA LUDOVICIANA (WILS.) BP.
  Louisiana Tanager.

✓111. HIRUNDO HORREORUM BARTON.
  Barn Swallow.

✓112. TACHYCINETA BICOLOR (V.) Cs.
  White-bellied Swallow.

113. TACHYCINETA THALASSINA (Sw.) CAB.
  Violet-green Swallow.

✓114. PETROCHELIDON LUNIFRONS (SAY) CAB.
  Cliff Swallow; Eave Swallow.

✓115. COTYLE RIPARIA (L.) BOIE.
  Bank Swallow; Sand Martin.

116. STELGIDOPTERYX SERRIPENNIS (AUD.) BD.
  Rough-winged Swallow.

✓117. PROGNE PURPUREA (L.) BOIE.
  Purple Martin.

118. AMPELIS GARRULUS L.
  Bohemian Waxwing.

✓119. AMPELIS CEDRORUM (V.) BD.
  Cedar Bird; Cherry Bird.

GEN. 51–53 OF KEY. 25

120. PHÆNOPEPLA NITENS (Sw.) Scl.
Black Ptilogonys.

121. MYIADESTES TOWNSENDII (Aud.) Cab.
Townsend's Flycatching Thrush.

✓ 122. VIREO OLIVACEUS (L.) V.
Red-eyed Vireo.

✓ 123. VIREO ALTILOQUUS (V.) Gr.,
var. barbatulus (Cab.) Cs.
Black-whiskered Vireo.

124. VIREO PHILADELPHICUS Cass.
Brotherly-love Vireo.

✓ 125. VIREO GILVUS (V.) Bp.
Warbling Vireo.

✓ 125a. VIREO GILVUS (V.) Bp.
var. swainsoni (Bd.) Cs.
Western Warbling Vireo.

✓ 126. VIREO FLAVIFRONS V.
Yellow-throated Vireo.

✓ 127. VIREO SOLITARIUS (Wils.) V.
Blue-headed Vireo; Solitary Vireo.

127a. VIREO SOLITARIUS V.,
var. plumbeus (Cs.) All.
Plumbeous Vireo.

128? VIREO VICINIOR Coues.
Gray Vireo.

✓ 129. VIREO NOVEBORACENSIS (Gm.) Bp.
White-eyed Vireo.

130? VIREO HUTTONI Cass.
Hutton's Vireo.

✓131. **VIREO BELLII** Aud.
    Bell's Vireo.

132. **VIREO PUSILLUS** Coues.
    Least Vireo.

133. **VIREO ATRICAPILLUS** Woodh.
    Black-headed Vireo.

✓134. **COLLURIO BOREALIS** (V.) Bd.
    Great Northern Shrike; Butcherbird.

✓135. **COLLURIO LUDOVICIANUS** (L.) Bd.
    Loggerhead Shrike.

✓135a. **COLLURIO LUDOVICIANUS** (L.) Bd.,
    var. EXCUBITOROIDES (Sw.) Cs.
    White-rumped Shrike.

136. **HESPERIPHONA VESPERTINA** (Coop.) Bp.
    Evening Grosbeak.

137. **PINICOLA ENUCLEATOR** (L.) V.
    Pine Grosbeak.

[138.]? **PYRRHULA CASSINI** (Bd.) Tristr.
    Cassin's Bullfinch.

139. **CARPODACUS PURPUREUS** (Gm.) Gr.
    Purple Finch.

140. **CARPODACUS CASSINI** Bd.
    Cassin's Purple Finch.

141. **CARPODACUS FRONTALIS** (Say) Gr.
    Crimson-fronted Finch; House Finch.

141a.* **CARPODACUS FRONTALIS** (Say) Cab.,
    var. HÆMORRHOUS (Wagl.) Ridg.
    Mexican Purple Finch.

---

*Not in the Key. See Ridgway, Am. Jour. Sci. Art, v, p. 39.

142. **LOXIA LEUCOPTERA** (Wils.).
White-winged Crossbill.

143. **LOXIA CURVIROSTRA** L.,
var. AMERICANA (Wils.) Cs.
Common Crossbill.

143a. **LOXIA CURVIROSTRA** L.,
var. MEXICANA (Strickl.) Cs.
Large-billed Crossbill.

144. **LEUCOSTICTE TEPHROCOTIS** Sw.
Gray-crowned Finch.

144a. **LEUCOSTICTE TEPHROCOTIS** Sw.,
var. GRISEINUCHA (Brandt) Cs.
Gray-eared Finch.

145. **LEUCOSTICTE ARCTOA** (Pall.) Bp.
Siberian Finch.

146. **ÆGIOTHUS LINARIA** (L.) Cab.
Red-poll Linnet.

146a. **ÆGIOTHUS LINARIA** (L.) Cab.
var. FUSCESCENS Cs.
Dusky Red-poll.

146b. **ÆGIOTHUS LINARIA** (L.) Cab.
var. EXILIPES Cs.
American Mealy Red-poll.

[147.] **LINOTA FLAVIROSTRIS** (L.) Bp.
var. BREWSTERI (Ridg.) Cs.
Brewster's Linnet.

148. **CHRYSOMITRIS PINUS** (Wils.) Bp.
Pine Linnet.

149. **CHRYSOMITRIS TRISTIS** (L.) Bp.
American Goldfinch; Yellowbird.

150. **CHRYSOMITRIS LAWRENCEI** (Cass.) Bp.
Lawrence's Goldfinch.

151. **CHRYSOMITRIS PSALTRIA** (Say) Bp.
Arkansas Goldfinch.

151a. **CHRYSOMITRIS PSALTRIA** (Say) Bp.,
var. ARIZONÆ Cs.
Arizona Goldfinch.

151b. **CHRYSOMITRIS PSALTRIA** (Say) Bp.,
var. MEXICANA (Sw.) Cs.
Mexican Goldfinch.

152. **PLECTROPHANES NIVALIS** (L.) Meyer.
Snow Bunting.

153. **PLECTROPHANES LAPPONICUS** (L.) Selby.
Lapland Longspur.

154. **PLECTROPHANES PICTUS** Sw.
Painted Lark Bunting.

155. **PLECTROPHANES ORNATUS** Towns.
Chestnut-colored Lark Bunting.

156. **PLECTROPHANES MACCOWNII** Lawr.
McCown's Lark Bunting.

157? **CENTRONYX BAIRDII** (Aud.) Bd.
Baird's Bunting.

157bis.* **CENTRONYX OCHROCEPHALUS** Aiken.
Ochreous-headed Bunting.

158. **PASSERCULUS PRINCEPS** Maynard.
Maynard's Sparrow.

159. **PASSERCULUS SAVANNA** (Wils.) Bp.
Savanna Sparrow.

---

* Not in the Key. See Aiken, Am. Nat., vii, 237.

159a. **PASSERCULUS SAVANNA** (Wils.) Bp.,
var. ANTHINUS (Bp.) Cs.
Titlark Sparrow.

159b. **PASSERCULUS SAVANNA** (Wils.) Bp.,
var. SANDVICENSIS (Gm.) Cs.
Northwestern Sparrow.

160. **PASSERCULUS ROSTRATUS** (Cass.) Bd.
Beaked Sparrow.

160a. **PASSERCULUS ROSTRATUS** (Cass.) Bd.,
var. GUTTATUS (Lawr.) Cs.
St. Lucas Sparrow.

161. **POOECETES GRAMINEUS** (Gm.) Bd.
Bay-winged Bunting; Grass Finch.

161a. **POOECETES GRAMINEUS** (Gm.) Bd.,
var. CONFINIS Bd.
Western Grass Finch.

162. **COTURNICULUS PASSERINUS** (Wils.) Bp.
Yellow-winged Sparrow.

162a. **COTURNICULUS PASSERINUS** (Wils.) Bp.,
var. PERPALLIDUS Ridg.
Bleached Yellow-winged Sparrow.

163. **COTURNICULUS HENSLOWI** (Aud.) Bp.
Henslow's Sparrow.

164. **COTURNICULUS LECONTEI** (Aud.) Bp.
LeConte's Sparrow.

165. **AMMODROMUS MARITIMUS** (Wils.) Sw.
Seaside Finch.

166. **AMMODROMUS CAUDACUTUS** (Gm.) Sw.
Sharp-tailed Finch.

167. **MELOSPIZA LINCOLNI** (Aud.) Bd.
Lincoln's Finch.

168. **MELOSPIZA PALUSTRIS** (Wils.) Bd.
Swamp Sparrow.

169. **MELOSPIZA MELODIA** (Wils.) Bd.
Song Sparrow.

169a. **MELOSPIZA MELODIA** (Wils.) Bd.,
var. FALLAX (Bd.) Ridg.
Gray Song Sparrow.

169b. **MELOSPIZA MELODIA** (Wils.) Bd.,
var. GUTTATA (Nutt.) Ridg.
Cinereous Song Sparrow.

169c. **MELOSPIZA MELODIA** (Wils.) Bd.,
var. RUFINA (Brandt.) Ridg.
Rufous Song Sparrow.

169d. **MELOSPIZA MELODIA** (Wils.) Bd.,
var. HEERMANNI (Bd.) Ridg.
Heermann's Song Sparrow.

169e. **MELOSPIZA MELODIA** (Wils.) Bd.,
var. GOULDII (Bd.) Ridg.
Gould's Song Sparrow.

169f. **MELOSPIZA MELODIA** (Wils.) Bd.,
var. INSIGNIS (Bd.) Ridg.
Bischoff's Song Sparrow.

170. **PEUCÆA ÆSTIVALIS** (Licht.) Cab.
Bachman's Finch.

170a. **PEUCÆA ÆSTIVALIS** (Licht.) Cab.,
var. CASSINI (Woodh.) All.
Cassin's Finch.

171. **PEUCÆA RUFICEPS** (Cass.) Bd.
Rufous-crowned Finch.

171bis.* **PEUCÆA CARPALIS** Coues.
Rufous-winged Sparrow.

---

*Not in the Key. See Am. Nat., vii, p. 322.

172. **POOSPIZA BILINEATA** (Cass.) Scl.
   Black-throated Finch.

173. **POOSPIZA BELLI** (Cass.) Scl.
   Bell's Finch.

174. **JUNCO HYEMALIS** (L.) Scl.
   Snowbird.

175? **JUNCO OREGONUS** (Towns.) Scl.
   Oregon Snowbird.

176? **JUNCO CINEREUS** (Sw.) Cab.,
   var. caniceps (Woodh.) Cs.
   Cinereous Snowbird.

177. **SPIZELLA MONTICOLA** (Gm.) Bd.
   Tree Sparrow.

178. **SPIZELLA SOCIALIS** (Wils.) Bp.
   Chipping Sparrow.

178a. **SPIZELLA SOCIALIS** (Wils.) Bp.,
   var. arizonæ Cs.
   Arizona Chipping Sparrow.

179. **SPIZELLA PUSILLA** (Wils.) Bp.
   Field Sparrow.

180. **SPIZELLA PALLIDA** (Sw.) Bp.
   Clay-colored Sparrow.

180a. **SPIZELLA PALLIDA** (Sw.) Bp.,
   var. breweri (Cass.) Cs.
   Brewer's Sparrow.

181. **SPIZELLA ATRIGULARIS** (Cab.) Bd.
   Black-chinned Sparrow.

182. **ZONOTRICHIA ALBICOLLIS** (Gm.) Bp.
   White-throated Sparrow.

GEN. 74–80 OF KEY. 39

183. **ZONOTRICHIA LEUCOPHRYS** (Forst.) Sw.
White-crowned Sparrow.

183a. **ZONOTRICHIA LEUCOPHRYS** (Forst.) Sw.,
var. Gambeli (Nutt.) All.
Gambel's Sparrow.

184. **ZONOTRICHIA CORONATA** (Pall.) Bd.
Golden-crowned Sparrow.

185. **ZONOTRICHIA QUERULA** (Nutt.) Gamb.
Harris's Sparrow.

186. **CHONDESTES GRAMMACA** (Say) Bp.
Lark Finch.

[187.] **PASSER DOMESTICUS** Linn.
English Sparrow.

188. **PASSERELLA ILIACA** (Merrem.) Sw.
Fox Sparrow.

189. **PASSERELLA TOWNSENDII** (Aud.) Nutt.
Townsend's Fox Sparrow.

189a. **PASSERELLA TOWNSENDII** (Aud.) Nutt.,
var. schistacea (Bd.) Cs.
Slate-colored Fox Sparrow.

190. **CALAMOSPIZA BICOLOR** (Towns.) Bp.
Lark Bunting; White-winged Blackbird.

191. **EUSPIZA AMERICANA** (Gm.) Bp.
Black-throated Bunting.

192? **EUSPIZA TOWNSENDII** (Aud.) Bp.
Townsend's Bunting.

193. **GONIAPHEA LUDOVICIANA** (L.) Bowdich.
Rose-breasted Grosbeak.

194. **GONIAPHEA MELANOCEPHALA** (Sw.) —.
Black-headed Grosbeak.

195. **GONIAPHEA CÆRULEA** (L.).
   Blue Grosbeak.

196. **CYANOSPIZA CIRIS** (L.) Bd.
   Painted Finch; Nonpareil.

197. **CYANOSPIZA VERSICOLOR** (Bp.) Bd.
   Western Nonpareil.

198. **CYANOSPIZA AMŒNA** (Say) Bd.
   Lazuli Finch.

199. **CYANOSPIZA CYANEA** (L.) Bd.
   Indigo Bird.

[200.] **SPERMOPHILA MORELETII** Pucheran.
   Morelet's Finch.

[201.] **PHONIPARA BICOLOR** (L.) Bp.
   Black-faced Finch.

202. **PYRRHULOXIA SINUATA** Bp.
   Texas Cardinal.

203. **CARDINALIS VIRGINIANUS** (Brisson) Bp.
   Cardinal Redbird.

203a. **CARDINALIS VIRGINIANUS** (Brisson) Bp.,
   var. igneus (Bd.) Cs.
   Fiery Redbird.

204. **PIPILO ERYTHROPHTHALMUS** (L.) V.
   Towhee Bunting; Chewink.

204a. **PIPILO ERYTHROPHTHALMUS** (L.) V.,
   var. alleni Cs.
   White-eyed Towhee.

205. **PIPILO MACULATUS** Sw.,
   var. oregonus (Bell) Cs.
   Oregon Towhee.

205a. **PIPILO MACULATUS** Sw.,
    var. ARCTICUS (Sw.) Cs.
    Arctic Towhee.

205b. **PIPILO MACULATUS** Sw.,
    var. MEGALONYX (Bd.) Cs.
    Spurred Towhee.

206. **PIPILO FUSCUS** Sw.
Brown Towhee; Canon Finch.

206a. **PIPILO FUSCUS** Sw.,
    var. ALBIGULA (Bd.) Cs.
White-throated Towhee.

206b. **PIPILO FUSCUS** Sw.,
    var. CRISSALIS (Vig.) Cs.
    Crissal Towhee.

207. **PIPILO ABERTI** Bd.
    Abert's Towhee.

208. **PIPILO CHLORURUS** (Towns.) Bd.
    Green-tailed Towhee.

209. **EMBERNAGRA RUFIVIRGATA** Lawr.
    Green Finch.

210. **DOLICHONYX ORYZIVORUS** (L.) Sw.
    Bobolink; Reedbird; Ricebird.

211.* **MOLOTHRUS PECORIS** (Gm.) Sw.
    Cowbird.

211a. **MOLOTHRUS PECORIS** (Gm.) Sw.,
    var. OBSCURUS (Gm.) Cs.
    Dwarf Cowbird.

212. **AGELÆUS PHŒNICEUS** (L.) V.
    Red-winged Blackbird.

---

*This should stand as *Molothrus ater* (Gm.) Gr.

212a. **AGELÆUS PHŒNICEUS** (L.) V.,
var. GUBERNATOR (Wagl.) Cs.
Red-shouldered Blackbird.

212b. **AGELÆUS PHŒNICEUS** (L.) V.,
var. TRICOLOR (Nutt.) Cs.
Red-and-white-shouldered Blackbird.

213. **XANTHOCEPHALUS ICTEROCEPHALUS** (Bp.) Bd.
Yellow-headed Blackbird.

214. **STURNELLA MAGNA** (L.) Sw.
Fieldlark; Meadowlark.

214a. **STURNELLA MAGNA** (L.) Sw.,
var. NEGLECTA (Aud.) All.
Western Fieldlark.

215. **ICTERUS SPURIUS** (L.) Bp.
Orchard Oriole.

215a. **ICTERUS SPURIUS** (L.) Bp.,
var. AFFINIS (Lawr.) Cs.
Texan Orchard Oriole.

216. **ICTERUS BALTIMORE** (L.) DANDIN.
Baltimore Oriole.

✓ 217. **ICTERUS BULLOCKII** (Sw.) Bp.
Bullock's Oriole.

218. **ICTERUS CUCULLATUS** Sw.
Hooded Oriole.

219. **ICTERUS PARISORUM** Bp.
Scott's Oriole.

220. **ICTERUS MELANOCEPHALUS** (WAGL.) GR.,
var. AUDUBONII (Girand.) Cs.
Audubon's Oriole.

221. **SCOLECOPHAGUS FERRUGINEUS** (GM.) Sw.
Rusty Grackle.

222. **SCOLECOPHAGUS CYANOCEPHALUS** (Wagl.) Cab.
Blue-headed Grackle.

223. **QUISCALUS MACROURUS** Sw.
Great-tailed Grackle.

224. **QUISCALUS MAJOR** Vieil.
Boat-tailed Grackle; Jackdaw.

225. **QUISCALUS PURPUREUS** (Bartr.) Licht.
Purple Grackle; Crow Blackbird.

225a. **QUISCALUS PURPUREUS** (Bartr.) Licht.,
var. aglæus (Bd.) Cs.
Florida Grackle.

226. **CORVUS CORAX** Linn.
Raven.

227. **CORVUS CRYPTOLEUCUS** Couch.
White-necked Raven.

228. **CORVUS AMERICANUS** Aud.
Common Crow.

228a. **CORVUS AMERICANUS** Aud.,
var. floridanus Bd.
Florida Crow.

228b. **CORVUS AMERICANUS** Aud.,
var. caurinus (Bd.) Cs.
Northwestern Fish Crow.

229. **CORVUS OSSIFRAGUS** Wils.
Fish Crow.

230. **PICICORVUS COLUMBIANUS** (Wils.) Bp.
Clarke's Crow.

231. **GYMNOKITTA CYANOCEPHALA** Maxim.
Blue Crow.

232. **PSILORHINUS MORIO** (Wagl.) Gr.
      Brown Jay.

233. **PICA MELANOLEUCA** V.,
      var. HUDSONICA (Sab.) All.
      American Magpie.

233a. **PICA MELANOLEUCA** V.,
      var. NUTTALLI (Aud.) Cs.
      Yellow-billed Magpie.

234. **CYANURUS CRISTATUS** (L.) Sw.
      Blue Jay.

235. **CYANURUS STELLERI** (Gm.) Sw.
      Steller's Jay.

235a. **CYANURUS STELLERI** (Gm.) Sw.,
      var. MACROLOPHA (Bd.) All.
      Long-crested Jay.

235b.* **CYANURUS STELLERI** (Gm.) Sw.,
      var. FRONTALIS Ridg.
      Blue-fronted Jay.

236. **APHELOCOMA FLORIDANA** (Bartram) Cab.
      Florida Jay.

236a. **APHELOCOMA FLORIDANA** (Bartr.) Cab.,
      var. WOODHOUSEI (Bd.) All.
      Woodhouse's Jay.

236b. **APHELOCOMA FLORIDANA** (Bartr.) Cab.,
      var. CALIFORNICA (Vig.) Cs.
      Californian Jay.

237. **APHELOCOMA SORDIDA** (Sw.) Cab.
      Sieber's Jay.

---

*Not in the Key. See Ridgway, Am. Journ., v, p. 43.

238. **XANTHOURA YNCAS** (Bodd.) Bp.,
  var. luxuosa (Less.) Cs.
  Rio Grande Jay.

239. **PERISOREUS CANADENSIS** (L.) Bp.
  Canada Jay.

[240.] **MILVULUS TYRANNUS** (L.) Bp.
  Fork-tailed Flycatcher.

241. **MILVULUS FORFICATUS** (Gm.) Sw.
  Swallow-tailed Flycatcher.

242. **TYRANNUS CAROLINENSIS** (L.) Bd.
  Kingbird; Bee-martin.

243. **TYRANNUS DOMINICENSIS** (Gm.) Rich.
  Gray Kingbird.

244. **TYRANNUS VERTICALIS** Say.
  Arkansas Flycatcher.

245. **TYRANNUS VOCIFERANS** Sw.
  Cassin's Flycatcher.

[246.] **TYRANNUS MELANCHOLICUS** V.,
  var. couchii (Bd.) Cs.
  Couch's Flycatcher.

247. **MYIARCHUS CRINITUS** (L.) Cab.
  Great-crested Flycatcher.

248. **MYIARCHUS CINERASCENS** Lawr.
  Ash-throated Flycatcher.

[249.] **MYIARCHUS LAWRENCEI** (Giraud.) Bd.
  Lawrence's Flycatcher.

250. **SAYORNIS SAYUS** (Bp.) Bd.
  Say's Flycatcher.

251. **SAYORNIS NIGRICANS** (Sw.) Bp.
  Black Flycatcher.

GEN. 107–110 OF KEY. 53

252. **SAYORNIS FUSCUS** (GM.) BD.
    Pewee; Pewit; Phœbe.

253. **CONTOPUS BOREALIS** (SW.) BD.
    Olive-sided Flycatcher.

254. **CONTOPUS PERTINAX** CAB.
    Coues' Flycatcher.

255. **CONTOPUS VIRENS** (L.) CAB.
    Wood Pewee.

255a. **CONTOPUS VIRENS** (L.) CAB.,
    var. RICHARDSONII (SW.) Cs.,
    Western Wood Pewee.

256. **EMPIDONAX ACADICUS** (GM.) BD.
    Acadian Flycatcher.

257. **EMPIDONAX TRAILLII** (AUD.) BD.
    Traill's Flycatcher.

257a. **EMPIDONAX TRAILLII** (AUD.) BD.,
    var. PUSILLUS (Bd.) Cs.
    Little Western Flycatcher.

258. **EMPIDONAX MINIMUS** BD.
    Least Flycatcher.

259. **EMPIDONAX FLAVIVENTRIS** BD.
    Yellow-bellied Flycatcher.

260. **EMPIDONAX HAMMONDII** BD.
    Hammond's Flycatcher.

261. **EMPIDONAX OBSCURUS** (SW.) BD.
    Wright's Flycatcher.

262. **MITREPHORUS FULVIFRONS** (GIRAUD.) SCL.,
    var. PALLESCENS Cs.
    Buff-breasted Flycatcher.

263. **PYROCEPHALUS RUBINEUS** (Bodd.) Gr.,
var. mexicanus (Scl.) Cs.
Vermilion Flycatcher.

264. **ANTROSTOMUS CAROLINENSIS** (Gm.) Gould.
Chuck-will's-widow.

265. **ANTROSTOMUS VOCIFERUS** (Wils.) Bp.
Whippoorwill; Night-jar.

266. **ANTROSTOMUS NUTTALLII** (Aud.) Cass.
Nuttall's Whippoorwill.

267. **CHORDEILES VIRGINIANUS** (Briss.) Bp.
Nighthawk.

267a. **CHORDEILES VIRGINIANUS** (Briss.) Bp.,
var. henryi (Cass.) All.
Western Nighthawk.

268. **CHORDEILES TEXENSIS** Lawr.
Texas Nighthawk.

269. **PANYPTILA SAXATILIS** (Woodh.) Cs.
White-throated Swift.

270. **NEPHŒCETES NIGER** (Gm.) Bd.,
var. borealis (Kennerly) Cs.
Black Swift.

271. **CHÆTURA PELASGIA** (L.) Steph.
Chimney Swift.

272? **CHÆTURA VAUXII** (Towns.) DeKay.
Vaux's Swift.

273. **HELIOPÆDICA XANTUSII** Lawr.
Xantus Hummingbird.

[274.] **LAMPORNIS MANGO** (L.) Sw.,
(*var.* porphyrula?)
Black-throated Hummingbird.

275. **TROCHILUS COLUBRIS** L.
   Ruby-throated Hummingbird.

276. **TROCHILUS ALEXANDRI** Bourc.
   Black-chinned Hummingbird.

277. **SELASPHORUS RUFUS** (Gm.) Sw.
   Rufous-backed Hummingbird.

278. **SELASPHORUS PLATYCERCUS** (Sw.) Gld.
   Broad-tailed Hummingbird.

279. **SELASPHORUS ANNA** (Less.)—.
   Anna Hummingbird.

280. **SELASPHORUS COSTÆ** (Bourc.) Bp.
   Costa Hummingbird.

281. **SELASPHORUS HELOISÆ** (——) ——.
   Heloise Hummingbird.

282. **STELLULA CALLIOPE** (——) Gld.
   Calliope Hummingbird.

[283]. **AGYRTRIA LINNÆI** (Bp.) ——.
   Linne Hummingbird.

[284]. **TROGON MEXICANUS** Sw.
   Mexican Trogon.

[285]. **MOMOTUS CÆRULEICEPS** Gould.
   Blue-headed Sawbill.

286. **CERYLE ALCYON** (L.) Boie.
   Belted Kingfisher.

287. **CERYLE AMERICANA** (Gm.) Boie,
   var. cabanisi (Reich.) Cs.
   Cabanis' Kingfisher.

288. **CROTOPHAGA ANI** L.
   Ani.

289. **GEOCOCCYX CALIFORNIANUS** (Less.) Bd.
   Ground Cuckoo; Chaparral Cock.

GEN. 128–131 OF KEY. 59

290. **COCCYZUS ERYTHROPHTHALMUS** (Wils.) Bd.
  Black-billed Cuckoo.

291. **COCCYZUS AMERICANUS** (L.) Bp.
  Yellow-billed Cuckoo.

292. **COCCYZUS SENICULUS** (Lath.)——.
  Mangrove Cuckoo.

293. **CAMPEPHILUS PRINCIPALIS** (L.) Gr.
  Ivory-billed Woodpecker.

294. **HYLOTOMUS PILEATUS** (L.) Bd.
  Pileated Woodpecker; Logcock.

295. **PICUS ALBOLARVATUS** (Cass.) Bd.
  White-headed Woodpecker.

296. **PICUS BOREALIS** V.
  Red-cockaded Woodpecker.

297. **PICUS SCALARIS** Wagler.
  Texas Woodpecker.

297a. **PICUS SCALARIS** Wagl.,
  var. NUTTALLI (Gamb.) Cs.
  Nuttall's Woodpecker.

297b. **PICUS SCALARIS** Wagl.,
  var. LUCASANUS (Xant.) Cs.
  St. Lucas Woodpecker.

298. **PICUS VILLOSUS** L.
  Hairy Woodpecker.

298a. **PICUS VILLOSUS** L.,
  var. HARRISI (Aud.) All.
  Harris' Woodpecker.

299. **PICUS PUBESCENS** L.
  Downy Woodpecker.

299a. **PICUS PUBESCENS** L.,
   var. GAIRDNERII (Aud.) Cs.
   Gairdner's Woodpecker.

300. **PICOIDES ARCTICUS** (Sw.) Gr.
   Black-backed Woodpecker.

301. **PICOIDES AMERICANUS** Brehm.
   Banded-backed Woodpecker.

301a. **PICOIDES AMERICANUS** Brehm.,
   var. DORSALIS (Bd.) All.
   Striped-backed Woodpecker.

302. **SPHYRAPICUS VARIUS** (L.) Bd.
   Yellow-bellied Woodpecker.

302a. **SPHYRAPICUS VARIUS** (L.) Bd.,
   var. NUCHALIS (Bd.) All.
   Nuchal Woodpecker.

303?* **SPHYRAPICUS RUBER** (Gm.) Bd.
   Red-breasted Woodpecker.

304. **SPHYRAPICUS THYROIDEUS** (Cass.) Bd.
   Brown-headed Woodpecker.

305. **SPHYRAPICUS WILLIAMSONI** (Newb.) Bd.
   Williamson's Woodpecker.

306. **CENTURUS CAROLINUS** (L.) Bp.
   Red-bellied Woodpecker.

307. **CENTURUS AURIFRONS** (Wagl.).
   Yellow-faced Woodpecker.

308. **CENTURUS UROPYGIALIS** Bd.
   Gila Woodpecker.

---

* Apparently a var. of 302.

309. **MELANERPES ERYTHROCEPHALUS** (L.) Sw.
Red-headed Woodpecker.

310. **MELANERPES FORMICIVORUS** (Sw.) Bp.
Californian Woodpecker.

310a. **MELANERPES FORMICIVORUS** (Sw.) Bp.,
var. ANGUSTIFRONS Bd.
Narrow-fronted Woodpecker.

311. **ASYNDESMUS TORQUATUS** (Wils.) Cs.
Lewis' Woodpecker.

312. **COLAPTES AURATUS** (L.) Sw.
Golden-winged Woodpecker; Flicker.

313. **COLAPTES CHRYSOIDES** Malh.
Gilded Woodpecker.

314. **COLAPTES MEXICANUS** Sw.
Red-shafted Woodpecker.

315. **CONURUS CAROLINENSIS** (L.) Kuhl.
Carolina Parroquet.

316. **STRIX FLAMMEA** L.,
var. AMERICANA (Aud.) Cs.
Barn Owl.

317. **BUBO VIRGINIANUS** (Gm.) Bp.
Great Horned Owl.

317a. **BUBO VIRGINIANUS** (Gm.) Bp.,
var. ARCTICUS (Sw.) Cass.
Arctic Horned Owl.

317b. **BUBO VIRGINIANUS** (Gm.) Bp.,
var. PACIFICUS Cass.
Pacific Horned Owl.

318. **SCOPS ASIO** (L.) Bp.
Screech Owl; Mottled Owl.

318a. **SCOPS ASIO** (L.) Bp.,
 var. KENNICOTTII (Ell.) Cs.
Kennicott's Owl.

318b. **SCOPS ASIO** (L.) Bp.,
 var. MACCALLII (Cass.) Cs.
McCall's Owl.

319. **SCOPS FLAMMEOLA** Scl.
Flammulated Owl.

320. **OTUS VULGARIS** (L.),
 var. WILSONIANUS (Less.) All.
Long-eared Owl.

321. **BRACHYOTUS PALUSTRIS** Auct.
Short-eared Owl.

322. **SYRNIUM LAPPONICUM** (L.),
 var. CINEREUM (Gm.) Ridg.
Great Gray Owl.

323. **SYRNIUM NEBULOSUM** (Forst.) Gr.
Barred Owl.

324. **SYRNIUM OCCIDENTALE** Xant.
Western Barred Owl.

325. **NYCTEA NIVEA** (Daud.) Gr.
Snowy Owl.

326. **SURNIA ULULA** (L.) Bp.,
 var. HUDSONICA (Gm.) Ridg.
Hawk Owl; Day Owl.

327. **NYCTALE TENGMALMI** (Gm.),
 var. RICHARDSONII (Bp.) Ridg.
Tengmalm's Owl.

328. **NYCTALE ACADICA** (Gm.) Bp.
Acadian Owl; Saw-whet Owl.

329. **GLAUCIDIUM PASSERINUM,**
var. CALIFORNICUM (Scl.) Ridg.
Pygmy Owl.

330. **GLAUCIDIUM FERRUGINEUM.**
Ferrugineous Owl.

331. **MICRATHENE WHITNEYI** (Coop.) Cs.
Whitney's Owl.

332. **SPEOTYTO CUNICULARIA** (Mol.),
var. HYPOGÆA (Bp.) Cs.
Burrowing Owl.

333. **CIRCUS CYANEUS** (L.) Lacép.,
var. HUDSONIUS (L.) Cs.
Marsh Hawk; Harrier.

334. **ROSTRHAMUS SOCIABILIS** (V.) D'Orb.
Everglade Kite.

335. **ICTINIA MISSISSIPPIENSIS** (Wils.) Gr.
Mississippi Kite.

336. **ELANUS LEUCURUS** (V.) Bp.
White-tailed Kite; Black-shouldered Kite.

337. **NAUCLERUS FURCATUS** (L.) Vig.
Swallow-tailed Kite.

338. **ACCIPITER FUSCUS** (Gm.) Bp.
Sharp-shinned Hawk; Pigeon Hawk.

339. **ACCIPITER COOPERI** Bp.
Cooper's Hawk; Chicken Hawk.

340. **ASTUR ATRICAPILLUS** (Wils.) Bp.
Goshawk.

341. **FALCO SACER** Forst.
Gyrfalcon; Jerfalcon.

341a. **FALCO SACER** Forst.,
var. CANDICANS (Gm.) Ridg.
Greenland Gyrfalcon.

342. **FALCO MEXICANUS** Licht.
Lanier Falcon.

343. **FALCO COMMUNIS** Variorum.
Peregrine Falcon; Duck Hawk.

344. **FALCO COLUMBARIUS** L.
Pigeon Hawk.

345. **FALCO RICHARDSONII** Ridg.
Richardson's Falcon.

346. **FALCO SPARVERIUS** L.
Sparrow Hawk.

346a. **FALCO SPARVERIUS** L.,
var. ISABELLINUS (Sw.) Ridg.
Isabella Sparrow Hawk.

347. **FALCO FEMORALIS** Temm.
Femoral Falcon.

348. **BUTEO UNICINCTUS** (Temm.) Gr.,
var. HARRISI (Aud.) Ridg.
Harris' Buzzard.

349? **BUTEO COOPERI** Cass.
Cooper's Buzzard.

350? **BUTEO HARLANI** (Aud.) Bp.
Harlan's Buzzard.

351. **BUTEO BOREALIS** (Gm.) V.
Red-tailed Buzzard; Hen Hawk.

351a. **BUTEO BOREALIS** (Gm.) V.,
   var. CALURUS (Cass.) Ridg.
Western Red-tailed Buzzard.

351b. **BUTEO BOREALIS** (Gm.) V.,
   var. LUCASANUS Ridg.
St. Lucas Buzzard.

351c.* **BUTEO BOREALIS** (Gm.) V.,
   var. KRIDERI.
Krider's Buzzard.

352. **BUTEO LINEATUS** (Gm.) Jard.
Red-shouldered Buzzard.

352a. **BUTEO LINEATUS** (Gm.) Jard.,
   var. ELEGANS (Cass) Ridg.
Western Red-shouldered Buzzard.

353. **BUTEO ZONOCERCUS** Scl.
Band-tailed Hawk.

354. **BUTEO SWAINSONI** Bp.
Swainson's Buzzard.

355. **BUTEO PENNSYLVANICUS** (Wils.) Bp.
Broad-winged Buzzard.

356. **ARCHIBUTEO LAGOPUS** (Brunn.) Gr.,
   var. SANCTI-JOHANNIS (Gm.) Ridg.
Rough-legged Buzzard.

357. **ARCHIBUTEO FERRUGINEUS** (Licht.) Gr.
Ferrugineous Buzzard.

358. **ASTURINA PLAGIATA** Schlegel.
Gray Hawk.

359.† **ONYCHOTES GRUBERI** Ridg.
Gruber's Buzzard.

---

* 351c. Not in Key; not published at date of going to press.
† 359. Questionably North American.

360. **PANDION HALIAETUS** (L.) SAVIGNY.
   Fish Hawk; Osprey.

361. **AQUILA CHRYSAETUS** (L.). *Cuv.*
   Golden Eagle.

362. **HALIAETUS LEUCOCEPHALUS** (L.) SAVIGNY.
   White-headed Eagle; Bald Eagle.

363. **POLYBORUS THARUS** (MOLL.) CASS.,
   var. AUDUBONII (Cass.) Ridg.
   Audubon's Caracara.

364. **CATHARTES CALIFORNIANUS** (SHAW) CUV.
   Californian Vulture.

365. **CATHARTES AURA** (L.) ILLIGER.
   Turkey Buzzard.

366. **CATHARTES ATRATUS** (BARTR.) LESS.
   Black Vulture; Carrion Crow.

367. **COLUMBA FASCIATA** SAY.
   Band-tailed Pigeon.

368. **COLUMBA FLAVIROSTRIS** WAGLER.
   Red-billed Pigeon.

369. **COLUMBA LEUCOCEPHALA** L.
   White-crowned Pigeon.

370. **ECTOPISTES MIGRATORIUS** (L.) SW.
   Wild Pigeon.

371. **ZENÆDURA CAROLINENSIS** (L.) BP.
   Carolina Dove.

372. **ZENÆDA AMABILIS** BP.
   Zenaida Dove.

373. **MELOPELEIA LEUCOPTERA** (L.) BP.
   White-winged Dove.

374. **CHAMÆPELEIA PASSERINA** (L.) Sw.
Ground Dove.

374a. **CHAMÆPELEIA PASSERINA** (L.) Sw.,
var. PALLESCENS (Bd.) Cs.
St. Lucas Ground Dove.

375. **SCARDAFELLA SQUAMOSA** (Temm.) Bp.,
var. INCA (Less.) Cs.
Scaled Dove.

376. **GEOTRYGON MARTINICA** (Gm.) Reich.
Key West Dove.

377. **STARNŒNAS CYANOCEPHALA** (L.) Bp.
Blue-headed Ground Dove.

378. **ORTALIDA VETULA** (Wagl.).
Texan Guan.

379. **MELEAGRIS GALLOPAVO** L.
Turkey.

379a. **MELEAGRIS GALLOPAVO** L.,
var. AMERICANA (Bartr.) Cs.
Common Wild Turkey.

380. **TETRAO CANADENSIS** L.
Canada Grouse; Spruce Partridge.

380a. **TETRAO CANADENSIS** L.,
var. FRANKLINI (Douglas) Cs.
Franklin's Grouse.

381. **TETRAO OBSCURUS** Say.
Dusky Grouse.

381a. **TETRAO OBSCURUS** Say,
var. RICHARDSONII (Dougl.) Cs.
Richardson's Grouse.

382. **CENTROCERCUS UROPHASIANUS** (Bp.) Sw.
Sage Cock; Cock-of-the-Plains.

383. **PEDIŒCETES PHASIANELLUS** (L.) Ell.
Northern Sharp-tailed Grouse.

383a. **PEDIŒCETES PHASIANELLUS** (L.) Ell.,
var. Columbianus (Ord.) Cs.
Common Sharp-tailed Grouse.

384. **CUPIDONIA CUPIDO** (L.) Bd.
Pinnated Grouse; Prairie Hen.

385. **BONASA UMBELLUS** (L.) Steph.
Ruffed Grouse; Partridge; Pheasant.

385a. **BONASA UMBELLUS** (L.) Steph.,
var. umbelloides (Dougl.) Bd.
Gray Ruffed Grouse.

385b. **BONASA UMBELLUS** (L.) Steph.,
var. sabinei (Dougl.) Cs.
Oregon Ruffed Grouse.

386. **LAGOPUS ALBUS** (Gm.) Aud.
Willow Ptarmigan.

387. **LAGOPUS RUPESTRIS** (Gm.) Leach.
Rock Ptarmigan.

388. **LAGOPUS LEUCURUS** Sw.
White-tailed Ptarmigan.

389. **ORTYX VIRGINIANUS** (L.) Bp.
Virginia Partridge; Quail; Bob-white.

389a. **ORTYX VIRGINIANUS** (L.) Bp.,
var. floridanus Cs.
Florida Partridge.

389b. **ORTYX VIRGINIANUS** (L.) Bp.,
var. TEXANUS (Lawr.) Cs.
Texan Partridge.

390. **OREORTYX PICTUS** (Dougl.) Bd.
Plumed Partridge.

391. **LOPHORTYX CALIFORNICUS** (Shaw) Bp.
Californian Partridge.

392. **LOPHORTYX GAMBELI** Nutt.
Gambel's Partridge.

393. **CALLIPEPLA SQUAMATA** (Vig.) Gr.
Scaled Partridge.

394. **CYRTONYX MASSENA** (Less.) Gld.
Massena Partridge.

395. **SQUATAROLA HELVETICA** (L.) Cuv.
Black-bellied Plover.

396. **CHARADRIUS FULVUS** Gm.,
var. VIRGINICUS (Borck.) Cs.
Golden Plover.

397. **ÆGIALITIS VOCIFERUS** (L.) Cass.
Kildeer Plover.

398. **ÆGIALITIS WILSONIUS** (Ord) Cass.
Wilson's Plover.

399. **ÆGIALITIS SEMIPALMATUS** (Bp.) Cab.
Semipalmated Plover; Ringneck.

400. **ÆGIALITIS MELODUS** (Ord) Cab.
Piping Plover; Ringneck.

401. **ÆGIALITIS CANTIANUS** (Lath.).
Snowy Plover.

402.* **ÆGIALITIS ASIATICUS** (Pall.),
   var. montanus (Towns.) Cs.
   Mountain Plover.

403. **APHRIZA VIRGATA** (Gm.) Gr.
   Surf Bird.

404. **HÆMATOPUS PALLIATUS** Temm.
   Oyster-catcher.

405. **HÆMATOPUS NIGER** Pallas.
   Black Oyster-catcher.

406. **STREPSILAS INTERPRES** (L.) Ill.
   Turnstone.

406a. **STREPSILAS INTERPRES** (L.) Ill.,
   var. melanocephalus (Vig.) Cs.
   Black-headed Turnstone.

407. **RECURVIROSTRA AMERICANA** Gm.
   Avocet.

408. **HIMANTOPUS NIGRICOLLIS** V.
   Stilt.

409. **STEGANOPUS WILSONI** (Sab.) Cs.
   Wilson's Phalarope.

410. **LOBIPES HYPERBOREUS** (L.) Cuv.
   Northern Phalarope.

411. **PHALAROPUS FULICARIUS** (L.) Bp.
   Red Phalarope.

412. **PHILOHELA MINOR** (Gm.) Gr.
   American Woodcock.

[413.] **SCOLOPAX RUSTICOLA** L.
   European Woodcock.

---

* May require to stand as *Eudromias montanus* (Towns.) Harting.

414. **GALLINAGO WILSONI** (Temm.) Bp.
American Snipe; Wilson's Snipe.

415. **MACRORHAMPHUS GRISEUS** (Gm.) Leach.
Red-breasted Snipe.

415a. **MACRORHAMPHUS GRISEUS** (Gm.) Leach,
var. scolopaceus (Say) Cs.
Long-billed Snipe.

416. **MICROPALAMA HIMANTOPUS** (Bp.) Bd.
Stilt Sandpiper.

417. **EREUNETES PUSILLUS** (L.) Cass.
Semipalmated Sandpiper.

417a. **EREUNETES PUSILLUS** (L.) Cass.,
var. occidentalis (Lawr.) Cs.
Western Semipalmated Sandpiper.

418. **TRINGA MINUTILLA** V.
Least Sandpiper.

419. **TRINGA BAIRDII** Coues.
Baird's Sandpiper.

420. **TRINGA MACULATA** V.
Pectoral Sandpiper.

421. **TRINGA BONAPARTEI** Schl.
White-rumped Sandpiper.

422? **TRINGA COOPERI** Bd.
Cooper's Sandpiper.

423. **TRINGA MARITIMA** Brunnich.
Purple Sandpiper.

424. **TRINGA ALPINA** L.,
var. americana Cass.
American Dunlin.

425. **TRINGA SUBARQUATA** Guld.
　　Curlew Sandpiper.

426. **TRINGA CANUTUS** L.
Red-breasted Sandpiper; Knot.

426bis.* **TRINGA CRASSIROSTRIS** Schlegel.
　　Thick-billed Sandpiper.

427. **CALIDRIS ARENARIA** (L.) Ill.
　　Sanderling; Ruddy Plover.

428. **LIMOSA FEDOA** (L.) Ord.
　　Great Marbled Godwit.

429. **LIMOSA HUDSONICA** (Lath.) Sw.
　　Hudsonian Godwit.

430. **LIMOSA UROPYGIALIS** Gould.
　　White-rumped Godwit.

431. **TOTANUS SEMIPALMATUS** Gm.
　　Semipalmated Tattler; Willet.

432. **TOTANUS MELANOLEUCUS** Gm.
　　Greater Tell-tale.

433. **TOTANUS FLAVIPES** Gm.
　　Yellow-shanks.

[434.] **TOTANUS CHLOROPUS** Nilsson.
　　Green-shanks.

435. **TOTANUS SOLITARIUS** Wils.
　　Solitary Tattler.

436. **TRINGOIDES MACULARIUS** (L.) Gr.
　　Spotted Sandpiper.

---

*Not in the Key. Obtained at St. Paul's Island, by H. W. Elliot. Identified by J. E. Harting. See Dall, Am. Nat., vii, Oct., 1873, p. 634.

[437.] PHILOMACHUS PUGNAX (L.) Gr.
Ruff; Reeve.

438. ACTITURUS BARTRAMIUS (Wils.) Bp.
Bartramian Sandpiper; Upland Plover.

439. TRYNGITES RUFESCENS (V.) Cab.
Buff-breasted Sandpiper.

440. HETEROSCELUS INCANUS (Gm.) Cs.
Wandering Tattler.

441. NUMENIUS LONGIROSTRIS Wils.
Long-billed Curlew.

442. NUMENIUS HUDSONICUS Lath.
Hudsonian Curlew.

443. NUMENIUS BOREALIS (Forst.) Lath.
Esquimaux Curlew.

444. TANTALUS LOCULATOR L.
Wood Ibis.

445. IBIS FALCINELLUS Auct.,
var. ordii (Bp.) All.
Glossy Ibis.

446. IBIS ALBA (L.) V.
White Ibis.

[447.] IBIS RUBRA (L.) V.
Scarlet Ibis.

448. PLATALEA AJAJA L.
Roseate Spoonbill.

449. ARDEA HERODIAS L.
Great Blue Heron.

450? ARDEA WURDEMANNI Bd.
Florida Heron.

451. **ARDEA OCCIDENTALIS** Aud.
Great White Heron.

✓ 452. **ARDEA EGRETTA** Gm.
Great White Egret.

✓ 453. **ARDEA CANDIDISSIMA** Jacquin.
Little White Egret.

454. **ARDEA LEUCOGASTRA** Gm.;
var. Leucoprymna (Licht.) Cs.
Louisiana Heron.

455. **ARDEA RUFA** Bodd.
Reddish Egret.

456. **ARDEA CÆRULEA** L.
Little Blue Heron.

✓ 457. **ARDEA VIRESCENS** L.
Green Heron.

✓ 458. **NYCTIARDEA GRISEA** (L.) Steph.,
var. Nævia (Bodd.) Allen.
Night Heron.

459. **NYCTIARDEA VIOLACEA** (L.) Sw.
Yellow-crowned Night Heron.

460. **BOTAURUS MINOR** (Gm.).
Bittern; Indian Hen.

461. **ARDETTA EXILIS** (Gm.) Gr.
Least Bittern.

462. **GRUS AMERICANUS** (L.) Ord.
White Crane; Whooping Crane.

463. **GRUS CANADENSIS** (L.) Temm.
Brown Crane; Sandhill Crane.

✓464. ARAMUS SCOLOPACEUS (Gm.) V.,
   var. GIGANTEUS (Bp.) Cs.
Scolopaceous Courlan.

✓465. RALLUS LONGIROSTRIS Bodd.
Clapper Rail; Salt-water Marsh Hen.

✓466. RALLUS ELEGANS Aud.
Fresh-water Marsh Hen.

ꞌ467. RALLUS VIRGINIANUS L.
Virginia Rail.

ꞌ468. PORZANA CAROLINA (L.) V.
Carolina Rail; Sora; Ortolan.

469. PORZANA NOVEBORACENSIS (Gm.) Cass.
Yellow Rail.

470. PORZANA JAMAICENSIS (Gm.) Cass.
Black Rail.

[471.] CREX PRATENSIS Bechstein.
Corn Crake.

✓472. GALLINULA GALEATA (Licht.) Bp.,
   (CHLOROPUS *var?*).
Florida Gallinule.

473. PORPHYRIO MARTINICA (L.) Temm.
Purple Gallinule.

474. FULICA AMERICANA Gm.
Coot.

475. PHŒNICOPTERUS RUBER L.
Flamingo.

476. CYGNUS BUCCINATOR Richardson.
Trumpeter Swan.

477. **CYGNUS AMERICANUS** Sharpless.
Whistling Swan.

478. **ANSER ALBIFRONS** Gm.,
var. Gambeli (Hartl.) Cs.
American White-fronted Goose.

479? **ANSER CÆRULESCENS** L.
Blue Goose.

480. **ANSER HYPERBOREUS** Pall.
Snow Goose.

480a. **ANSER HYPERBOREUS** Pall.,
var. albatus (Cass.) Cs.
Lesser Snow Goose.

481. **ANSER ROSSII** Bd.
Ross' Goose.

482. **PHILACTE CANAGICA** (Sevast.) Bann.
Painted Goose.

[483.] **BRANTA LEUCOPSIS** (L.).
Barnacle Goose.

484. **BRANTA BERNICLA** (L.).
Brant Goose.

485. **BRANTA CANADENSIS** (L.).
Canada Goose; Wild Goose.

485a. **BRANTA CANADENSIS** (L.),
var. leucopareia (Brandt) Cs.
White-collared Goose.

485b. **BRANTA CANADENSIS** (L.),
var. hutchinsii (Rich.) Cs.
Hutchins' Goose.

486. **DENDROCYGNA FULVA** (Gm.) Burm.
    Fulvous Tree Duck.

487. **DENDROCYGNA AUTUMNALIS** (L.) Eyton.
    Autumnal Tree Duck.

488. **ANAS BOSCHAS** L.
    Mallard.

489. **ANAS OBSCURA** Gm.
    Dusky Duck.

490. **DAFILA ACUTA** (L.) Jenyns.
    Pintail; Sprigtail.

491. **CHAULELASMUS STREPERUS** (L.) Gray.
    Gadwall; Gray Duck.

[492.] **MARECA PENELOPE** (L.) Bp.
    European Widgeon.

493? **MARECA AMERICANA** (Gm.) Steph.
    American Widgeon; Baldpate.

[494.] **QUERQUEDULA CRECCA** (L.) Steph.
    English Teal.

495. **QUERQUEDULA CAROLINENSIS** (Gm.).
    Green-winged Teal.

496. **QUERQUEDULA DISCORS** (L.) Steph.
    Blue-winged Teal.

497. **QUERQUEDULA CYANOPTERA** (V.) Cass.
    Cinnamon Teal.

498. **SPATULA CLYPEATA** (L.) Boie.
    Shoveller.

499. **AIX SPONSA** (L.) Boie.
    Summer Duck; Wood Duck.

500. **FULIGULA MARILA** (L.) Steph.
  Greater Blackhead.

501? **FULIGULA AFFINIS** Eyton.
  Lesser Blackhead.

502. **FULIGULA COLLARIS** (Donovan) Bp.
  Ring-necked Duck.

503. **FULIGULA FERINA** (L.) Sw.,
  var. americana (Eyton) Coues.
  Redhead; Pochard.

504. **FULIGULA VALLISNERIA** (Wils.) Steph.
  Canvas-back.

505. **BUCEPHALA CLANGULA** (L.) Gr.
  Golden-eyed Duck.

506. **BUCEPHALA ISLANDICA** (Gm.) Bd.
  Barrow's Golden-eye.

507. **BUCEPHALA ALBEOLA** (L.) Bd.
  Buffle-headed Duck.

508. **HARELDA GLACIALIS** (L.) Leach.
  Long-tailed Duck.

509. **CAMPTOLÆMUS LABRADORIUS** (Gm.) Gr.
  Labrador Duck.

510. **HISTRIONICUS TORQUATUS** (L.) Bp.
  Harlequin Duck.

511. **SOMATERIA STELLERI** (Pall.) Jardine.
  Steller's Duck.

512. **SOMATERIA FISCHERI** (Brandt) Coues.
  Spectacled Eider.

513. **SOMATERIA MOLLISSIMA** (L.) Leach.
  Eider Duck.

514? **SOMATERIA V-NIGRA** Gray.
 Pacific Eider.

515. **SOMATERIA SPECTABILIS** (L.) Leach.
 King Eider.

516. **ŒDEMIA AMERICANA** Sw.
 American Black Scoter.

517. **ŒDEMIA FUSCA** (L.) Sw.,
 (? *var.* velvetina Cass.)
 Velvet Scoter.

518. **ŒDEMIA PERSPICILLATA** (L.) Fleming.
 Surf Duck.

518a. **ŒDEMIA PERSPICILLATA** (L.) Fleming,
 var. trowbridgei (Bd.) Coues.
 Long-billed Scoter.

519. **ERISMATURA RUBIDA** (Wils.) Bp.
 Ruddy Duck.

[520.] **ERISMATURA DOMINICA** (L.) Eyton.
 St. Domingo Duck.

521. **MERGUS MERGANSER** L.
 Merganser; Goosander.

522. **MERGUS SERRATOR** L.
 Red-breasted Merganser.

523. **MERGUS CUCULLATUS** L.
 Hooded Merganser.

524. **SULA BASSANA** L.
 Gannet; Solan Goose.

525. **SULA FIBER** L.
 Booby Gannet.

526. **PELECANUS TRACHYRHYNCHUS** Lath.
 White Pelican.

527. **PELECANUS FUSCUS** L.
      Brown Pelican.

528. **GRACULUS CARBO** (L.) Gray.
      Common Cormorant; Shag.

529. **GRACULUS CINCINNATUS** (Brandt) Gray.
      White-tufted Cormorant.

530. **GRACULUS DILOPHUS** (Sw.) Gray.
      Double-crested Cormorant.

530a. **GRACULUS DILOPHUS** (Sw.) Gray,
      var. floridanus (Aud.) Coues.
      Florida Cormorant.

531. **GRACULUS MEXICANUS** (Brandt) Bp.
      Mexican Cormorant.

532. **GRACULUS PENICILLATUS** (Brandt) Bp.
      Brandt's Cormorant.

533. **GRACULUS PERSPICILLATUS** (Pall.) Lawr.
      Pallas' Cormorant.

534. **GRACULUS BICRISTATUS** (Pall.) Bd.
      Red-faced Cormorant.

535. **GRACULUS VIOLACEUS** (Gm.) Gr.
      Violet-green Cormorant.

536. **PLOTUS ANHINGA** L.
      Anhinga; Darter.

537. **TACHYPETES AQUILUS** (L.) V.
      Frigate.

538. **PHAETHON FLAVIROSTRIS** Brandt.
      Yellow-billed Tropic Bird.

539. **STERCORARIUS SKUA** (Brunn.) Coues.
      Skua Gull.

540. **STERCORARIUS POMATORHINUS** (Temm.) Lawr.
Pomarine Jaeger.

541. **STERCORARIUS PARASITICUS** (Brunn.) Gray.
Richardson's Jaeger.

542. **STERCORARIUS BUFFONI** (Boie) Coues.
Arctic Jaeger; Long-tailed Jaeger.

543. **LARUS GLAUCUS** Brunn.
Glaucous Gull.

544. **LARUS LEUCOPTERUS** Faber.
White-winged Gull.

545. **LARUS GLAUCESCENS** Licht.
Glaucous-winged Gull.

546. **LARUS MARINUS** L.
Great Black-backed Gull.

547. **LARUS ARGENTATUS** Brunn.
Herring Gull; Common Gull.

547a. **LARUS ARGENTATUS** Brunn.,
var. SMITHSONIANUS Coues.
American Herring Gull.

547b. **LARUS ARGENTATUS** Brunn.,
var. OCCIDENTALIS (Aud.) Coues.
Western Herring Gull.

548. **LARUS DELAWARENSIS** Ord.
Ring-billed Gull.

548a. **LARUS DELAWARENSIS** Ord,
var. CALIFORNICUS (Lawr.) Coues.
Californian Gull.

549. **LARUS CANUS** L.,
var. BRACHYRHYNCHUS (Rich.) Coues.
American Mew Gull.

550. **LARUS EBURNEUS** Gm.
Ivory Gull.

551. **LARUS BELCHERI** Vigors.
White-headed Gull

552. **LARUS TRIDACTYLUS** L.
Kittiwake Gull.

552a. **LARUS TRIDACTYLUS** L.,
var. kotzebui (Bp.) Coues.
Pacific Kittiwake.

553. **LARUS BREVIROSTRIS** (Brandt) Coues.
Short-billed Kittiwake.

554. **LARUS ATRICILLA** L.
Laughing Gull.

555. **LARUS FRANKLINI** Rich.
Franklin's Rosy Gull.

556. **LARUS PHILADELPHIA** (Ord) Coues.
Bonaparte's Gull.

557. **RHODOSTETHIA ROSEA** (Macgill.) Bp.
Wedge-tailed Gull.

558. **XEMA SABINEI** (Sab.) Bp.
Fork-tailed Gull.

559. **XEMA FURCATUM** (Neboux).
Swallow-tailed Gull.

560. **STERNA ANGLICA** Montagu.
Gull-billed Tern; Marsh Tern.

561. **STERNA CASPIA** Pallas,
var. imperator Coues.
Caspian Tern.

562. **STERNA REGIA** Gambel.
Royal Tern.

563. **STERNA GALERICULATA** Licht.
 Elegant Tern.

564. **STERNA CANTIACA** Gm.
 Sandwich Tern.

565. **STERNA HIRUNDO** L.
 Common Tern; Sea Swallow.

566. **STERNA FORSTERI** Nutt.
 Forster's Tern.

567. **STERNA MACROURA** Naumann.
 Arctic Tern.

568. **STERNA LONGIPENNIS** Nordmann.
 Pike's Tern.

569. **STERNA PARADISÆA** Brunn.
 Roseate Tern.

570. **STERNA SUPERCILIARIS** V.
 Least Tern.

[571.] **STERNA TRUDEAUI** Aud.
 Trudeau's Tern.

572. **STERNA ALEUTICA** Baird.
 Aleutian Tern.

573. **STERNA FULIGINOSA** Gm.
 Sooty Tern.

[574.] **STERNA ANOSTHÆTA** Scopoli.
 Bridled Tern.

575. **HYDROCHELIDON FISSIPES** (L.) Gray.
 Black Tern.

576. **ANOUS STOLIDUS** (L.) Leach.
 Noddy Tern.

577. **RHYNCHOPS NIGRA** L.
 Black Skimmer.

578. **DIOMEDEA BRACHYURA** Temm.
Short-tailed Albatross.

579. **DIOMEDEA NIGRIPES** Aud.
Black-footed Albatross.

580. **DIOMEDEA FULIGINOSA** Gm.
Sooty Albatross.

581. **FULMARUS GIGANTEUS** (Gm.).
Giant Fulmar.

582. **FULMARUS GLACIALIS** (L.) Steph.
Fulmar Petrel.

582a. **FULMARUS GLACIALIS** (L.) Steph.,
var. PACIFICUS (Aud.) Coues.
Pacific Fulmar.

582b. **FULMARUS GLACIALIS** (L.) Steph.,
var. RODGERSI (Cass.) Coues.
Rodgers' Fulmar.

[583.] **FULMARUS TENUIROSTRIS** (Aud.) Coues.
Slender-billed Fulmar.

[584.] **DAPTION CAPENSIS** (L.) Steph.
Pintado Petrel; Cape Pigeon.

[585.] **ÆSTRELATA HÆSITATA** (Kuhl) Coues.
Black-capped Petrel.

586. **HALOCYPTENA MICROSOMA** Coues.
Wedge-tailed Petrel; Least Petrel.

587. **PROCELLARIA PELAGICA** L.
Stormy Petrel; Mother Carey's Chicken.

588. **CYMOCHOREA LEUCORRHOA** (V.) Coues.
Leach's Petrel.

589. **CYMOCHOREA MELANIA** (Bp.) Coues.
Black Petrel.

590. **CYMOCHOREA HOMOCHROA** Coues.
Ashy Petrel.

591. **OCEANODROMA FURCATA** (Gm.) Bp.
Fork-tailed Petrel.

592. **OCEANODROMA HORNBYI** (Gray) Bp.
Hornby's Petrel.

593. **OCEANITES OCEANICA** (Kuhl) Coues.
Wilson's Petrel.

[594.] **FREGETTA GRALLARIA** (V.) Bp.
White-bellied Petrel.

[595.] **PUFFINUS MELANURUS** (Bonn.) Coues.
Black-tailed Shearwater.

596. **PUFFINUS KUHLII** Bp.
Cinereous Shearwater.

597. **PUFFINUS MAJOR** Faber.
Greater Shearwater.

598? **PUFFINUS CREATOPUS** Coues.
Flesh-footed Shearwater.

599. **PUFFINUS ANGLORUM** Temm.
Manks Shearwater.

600. **PUFFINUS OBSCURUS** (Gm.) Lath.
Dusky Shearwater.

601? **PUFFINUS OPISTHOMELAS** Coues.
Black-vented Shearwater.

602? **PUFFINUS FULIGINOSUS** Strickl.
Sooty Shearwater.

603? **PUFFINUS AMAUROSOMA** Coues.
Dark-bodied Shearwater.

604. **PUFFINUS TENUIROSTRIS** Temm.
Slender-billed Shearwater.

605. **COLYMBUS TORQUATUS** Brunn.
   Loon; Great Northern Diver.

605a. **COLYMBUS TORQUATUS** Brunn.,
   var. ADAMSII (Gray) Coues.
   Yellow-billed Loon.

606. **COLYMBUS ARCTICUS** L.
   Black-throated Diver.

606a. **COLYMBUS ARCTICUS** L.,
   var. PACIFICUS (Lawr.) Coues.
   Pacific Diver.

607. **COLYMBUS SEPTENTRIONALIS** L.
   Red-throated Diver.

608. **PODICEPS OCCIDENTALIS** Lawr.
   Western Grebe.

608a. **PODICEPS OCCIDENTALIS** Lawr.,
   var. CLARKII (Lawr.) Coues.
   Clarke's Grebe.

609. **PODICEPS CRISTATUS** (L.) Lath.
   Crested Grebe.

610. **PODICEPS GRISEIGENA** (Bodd.) Gray,
   var HOLBÖLLI (Reinh.) Coues.
   Red-necked Grebe.

611. **PODICEPS CORNUTUS** (Gm.) Lath.
   Horned Grebe.

612. **PODICEPS AURITUS** (L.) Lath.,
   var. CALIFORNICUS (Heerm.) Coues.
   American Eared Grebe.

613. **PODICEPS DOMINICUS** (L.)
   St. Domingo Grebe.

614. **PODILYMBUS PODICEPS** (L.) Lawr.
Pied-billed Dabchick.

615. **ALCA IMPENNIS** L.
Great Auk.
[Extinct ?]

616. **UTAMANIA TORDA** (L.) Leach.
Razor-billed Auk.

617. **FRATERCULA CORNICULATA** (Naum.) Gray.
Horned Puffin.

618. **FRATERCULA ARCTICA** (L.) Steph.
Common Puffin; Sea Parrot.

618a. **FRATERCULA ARCTICA** (L.) Steph.,
var. glacialis (Leach) Coues.
Large-billed Puffin.

619. **FRATERCULA CIRRHATA** (Pall.) Steph.
Tufted Puffin.

620. **CERATORHINA MONOCERATA** (Pall.) Cass.
Horn-billed Auk.

621. **PHALERIS PSITTACULA** (Pall.) Temm.
Parroquet Auk.

622. **SIMORHYNCHUS CRISTATELLUS** (Pall.) Merrem.
Crested Auk.

623. **SIMORHYNCHUS CAMTSCHATICUS** (Lepech.) Schl.
Whiskered Auk.

624. **SIMORHYNCHUS PUSILLUS** (Pall.) Coues.
Knob-billed Auk; Least Auk.

625. **PTYCHORHAMPHUS ALEUTICUS** (Pall.) Brandt.
Aleutian Auk.

626. **MERGULUS ALLE** (L.) Vieill.
  Sea Dove; Dovekie.

627. **SYNTHLIBORHAMPHUS ANTIQUUS** (Gm.) Brandt.
  Black-throated Guillemot.

628. **SYNTHLIBORHAMPHUS WURMIZUSUME** (Temm.) Coues.
  Temminck's Auk.

629. **BRACHYRHAMPHUS MARMORATUS** (Gm.) Brandt.
  Marbled Murrelet.

630. **BRACHYRHAMPHUS KITTLITZII** Brandt.
  Kittlitz's Murrelet.

631. **URIA GRYLLE** (L.) Brunn.
  Black Guillemot; Sea Pigeon.

632. **URIA COLUMBA** (Pall.) Cass.
  Pigeon Guillemot.

633. **URIA CARBO** (Pall.) Brandt.
  Sooty Guillemot.

634. **LOMVIA TROILE** (L.) Brandt.
  Common Guillemot; Murre.

635. **LOMVIA ARRA** (Pall.) Coues.
  Thick-billed Guillemot.

EXTINCT SPECIES.

1. **UINTORNIS LUCARIS** Marsh.

2. **AQUILA DANANA** Marsh.

3. **BUBO LEPTOSTEUS** Marsh.

4. **MELEAGRIS ANTIQUUS** Marsh.

5. **MELEAGRIS ALTUS** Marsh.

6. **MELEAGRIS CELER** Marsh.

7. **GRUS HAYDENI** Marsh.

8. **GRUS PROAVUS** Marsh.

9. **ALETORNIS NOBILIS** Marsh.

10. **ALETORNIS PERNIX** Marsh.

11. **ALETORNIS VENUSTUS** Marsh.

12. **ALETORNIS GRACILIS** Marsh.

13. **ALETORNIS BELLUS** Marsh.

14. **TELMATORNIS PRISCUS** Marsh.

15. **TELMATORNIS AFFINIS** Marsh.

EXTINCT SPECIES.

16. **PALÆOTRINGA LITTORALIS** Marsh.

17. **PALÆOTRINGA VETUS** Marsh.

18. **PALÆOTRINGA VAGANS** Marsh.

19. **SULA LOXOSTYLA** Cope.

20. **GRACULUS IDAHENSIS** Marsh.

21. **GRACULAVUS VELOX** Marsh.

22. **GRACULAVUS PUMILUS** Marsh.

23. **GRACULAVUS ANCEPS** Marsh.

23bis.* **GRACULAVUS AGILIS** Marsh.

24. **ICHTHYORNIS DISPAR** Marsh.

24bis.† **APATORNIS CELER** Marsh.

25. **PUFFINUS CONRADI** Marsh.

26. **CATARRACTES ANTIQUUS** Marsh.

27. **CATARRACTES AFFINIS** Marsh.

28. **HESPERORNIS REGALIS** Marsh.

29. **LAORNIS EDVARDSIANUS** Marsh.

---

*Not in the Key. (Marsh, Am. Jour., Sci. and Arts, v, p. 230, March, 1873.)

†Not in the Key. This species, with No. 24, represents a new order, *Ichthyornithes*, of a new subclass, *Odontornithes*. (Marsh, Am. Jour., Sci. and Arts, v, p. 161, Feb., 1873.)

APPENDIX

CONTAINING

ADDITIONS AND CORRECTIONS

TO THE CHECK LIST.

PREPARED under circumstances of remote isolation which deprived the author of the advantage of certain works of reference he desired to consult, the CHECK LIST contains some names for which no authority is cited, and in a few instances a change of the authority given may be required.

One new species has been added to the North American fauna during the printing of the List; five additional known species have since been ascertained to occur in this country, and meanwhile several new varieties have been published after the impression had passed the page where they should respectively appear; these are brought into the present connection. Most of them appear entitled to varietal recognition; but in printing the names formally, for the convenience of those who may desire to use such names in labelling, the author must not necessarily be held to endorse them in every instance.

The body of the List was printed, and some early copies distributed, in Dec., 1873; but the publication of the volume was held over until 1874, to insert in the Appendix names then about being published.

No. 41? The query indicates a probability that this is a variety of No. 40, as held by Mr. Allen.

No. 46. The United States form constitutes a variety of true *mexicanus*. See Ridgway, Am. Nat., vii, 1873, 603.

46. CATHERPES MEXICANUS (Sw.) BD.,
var. CONSPERSUS Ridg.
White-throated Wren.

No. 53a. The *pale* western Eremophila, not the same as the small bright southwestern var. *rufa*, may be distinguished as

53b. EREMOPHILA ALPESTRIS (FORST.),
var. LEUCOLÆMA Coues.
Prairie Lark.

No. 55bis. The following species, a straggler from Asia, is in the Smithsonian Institution from St. Michael's, Alaska, and should take place in the list.

[55bis.] ANTHUS PRATENSIS BECHST.
Meadow Pipit.

No. 68. The Pacific form is varietally distinguishable. See Ridgway, Am. Nat., vii, 1873, 606. The Floridan form, later distinguished by Mr. Ridgway as var. *obscurus*, seems hardly worthy of recognition by name.

68a. HELMINTHOPHAGA CELATA (SAY) BD.,
var. LUTESCENS Ridg.
Golden Orange-crowned Warbler.

No. 88. On the Mississippi Valley form, see Ridgway, Am. Nat., vii, 1873, 606.

88a. DENDRŒCA DOMINICA (L.) BD.,
var. ALBILORA Bd.
White-browed Yellow-throated Warbler.

No. 99? The probability mentioned in the text may be regarded as assured.
No. 102. The Pacific form is varietally distinguishable. See Ridgway, Am. Nat., vii, 1873, 608.

102a. MYIODIOCTES PUSILLUS (WILS.) BP.,
var. PILEOLATA (Pall.) Ridg.
Pacific Flycatching Warbler.

[No. 106.] According to Baird and Ridgway, Am. Nat., vii, 1873, 612, this should stand as *C. bahamensis*.

[106.] CERTHIOLA BAHAMENSIS Reich.
Honey Creeper.

No. 135a. The *C. elegans* of Baird (not of Swainson) has been renamed *C. ludovicianus* var. *robustus*, a name which, however, it may not be necessary to adopt. (Am. Nat., vii, 1873, 609.)

[No. 138]? The query indicates that the determination of specific validity, cited and adopted in the Key, may have been made by Dr. Tristram on grounds held in the Key to constitute only geographical varieties; so that we may revert to the view of its original describer as *P. coccinea* var. *cassini* Bd.

[138.] PYRRHULA COCCINEA,
var. CASSINI Bd.
Cassin's Bullfinch.

No. 144. The *Leucosticte tephrocotis* var. *australis* Allen, lately described by Mr. Ridgway (Ess. Inst. Bull., v, 197), I believe to be merely the midsummer plumage of the ordinary bird, as my *Ægiothus* var. *fuscescens* probably is of *A. linaria*.

No. 155. *For* chestnut-colored *read* chestnut-collared.

" 157. Omit the query, which should have been affixed to the next species.

No. 157bis. To be cancelled. See Scott, Am. Nat., vii, 1873, 564; Coues, *ibid.*, p. 696.

No. 165. There is a curious small blackish form of this species from Florida, which has been distinguished (Bull. Ess. Inst., v, 198) as

165a. AMMODROMUS MARITIMUS Sw.,
var. NIGRESCENS Ridgw.
Dusky Seaside Finch.

No. 170a. Mr. Ridgway has lately demonstrated to my satisfaction that *Peucæa cassini* is a distinct species; the bird which I called "var. *cassini*" is a variety of *æstivalis* which he proposes to call var. *arizonæ*. Am. Nat., vii, 1873, 616. So the species and varieties will stand:—

170a. PEUCÆA ÆSTIVALIS (LICHT.) CAB.,
var. ARIZONÆ Ridg.
Arizona Pine Finch.

170bis. PEUCÆA CASSINI (WOODH.) BD.
Cassin's Pine Finch.

No. 173. A very notable variety of *Poospiza belli*, from Nevada, has lately been characterized (Bull. Ess. Inst., v, 198). It is much larger, paler and grayer, with streaked interscapulars.

173a. POOSPIZA BELLI (CASS.) SCL.,
var. NEVADENSIS Ridg.
Nevadan Finch.

No. 174. The form of *Junco* with white wing-bars, noted in the Key, p. 141, is named *J. hyemalis* var. *aikeni* Ridgway, Am. Nat., vii, 1873, 616. See also Pr. Bost. Soc., xv, 1872, p. 201.

174a. JUNCO HYEMALIS (L.) Scl.,
var. AIKENI Ridg.
White-winged Snowbird.

Nos. 175? 176? The queries indicate the gradation with No. 174 noted in the Key, p. 141.

No. 183a. The true *Z. leucophrys* var. *gambeli* is a Pacific coast form, from which the Middle Coast form has been distinguished (Bull. Ess. Inst., v, 198) as

183b. ZONOTRICHIA LEUCOPHRYS (Forst.) Sw.,
var. INTERMEDIA Ridgw.
Ridgway's Sparrow.

No. 206. *For* Canon *read* Cañon.
" 210. The prairie form has been characterized as *Dolichonyx oryzivorus* var. *albinucha* Ridg. (Bull. Ess. Inst., v, 198), a name it may not be necessary to adopt.
No. 216. *For* Daudin *read* Daudin.
" 220. *For* Girand *read* Giraud.
" 226. *For* Nuttallii *read* Nuttalli.
" 229a. *For* Gairdnerii *read* Gairdneri.
" 237. The question of synonymy left open in the Key, p. 166, has been determined (Bull. Ess. Inst., v, 199) as follows: "*A. sordida*" of the Key is a new variety, *arizonæ* of *ultramarina*, the true *sordida* being a Mexican variety of the same species. Accordingly, No. 237 should stand as

237. APHELOCOMA ULTRAMARINA (Bp.) Cab.,
var. ARIZONÆ (Ridg.)
Arizona Ultramarine Jay.

No. 239. Two varieties of Canada jay, one from Alaska, the other from the Rocky Mountains, have lately been named (Bull. Ess. Inst., v, 199).

239a. PERISOREUS CANADENSIS (L.) Bp.,
var. OBSCURUS Ridg.
Dusky Canada Jay.

239b. PERISOREUS CANADENSIS (L.) Bp.,
var. CAPITALIS Bd.
Rocky Mountain Jay.

No. 274bis. A fine species of humming bird has been discovered by Mr. H. W. Henshaw to inhabit Arizona, and has been determined by Mr. Lawrence to be *Eugenes fulgens*. (Am. Nat., viii, 1874, in press.)

274bis. EUGENES FULGENS (Sw.).
Refulgent Hummingbird.

No. 279. The authority is (LESS.) BP.
" 281. The authority is (LESS.) GLD.
" 292. The authority is (LATH.) NUTT.
" 303? This species, queried in the text, and in Key, p. 195, may be regarded as a variety of 302, the intergradation, through 302a, proving complete. See Ridgway, Am. Jour., iv, Dec., 1872.

302b. SPHYRAPICUS VARIUS (L.) BD.,
var. RUBER (Gm.) Ridg.
Red-breasted Woodpecker.

Nos. 304, 305. Observations lately made by Mr. H. W. Henshaw (Am. Nat., viii, 1874, in press) are to the effect that *Sphyrapicus thyroideus* is the female of *S. williamsoni*. The opposite sexes of each of these species have not been satisfactorily recognized, and upon examination of Mr. Henshaw's material, I find almost conclusive evidences in favor of his views, substantiating his observations. Such sexual differences are unique in the family. As the older name, *S. thyroideus* will stand for the species, *S. williamsoni* becoming a synonyme. No. 305 is therefore to be cancelled.

No. 307. The authority is (WAGL.) GRAY.
No. 318. A dark Floridan form of *Scops* has been characterized under the following name (Bull. Ess. Inst., v, 200):—

318c. SCOPS ASIO (L.) BP.,
var. FLORIDANUS Ridg.
Floridan Screech Owl.

No. 320. The authority is FLEMING.
" 321. The authority is (BECHST.) BP.
" 322. The term *cinereum* has priority over *lapponicum;* the bird should stand as

322. SYRNIUM CINEREUM (GM.) AUD.,
Great Gray Owl.

No. 327. The authority is (GM.) BP.
" 329. The authority is (L.) BP.
" 330. The authority is (MAX.) KAUP.

Nos. 343, 344. The dark northwest coast forms of duck hawk and pigeon hawk respectively have been named (Bull. Ess. Inst., v, 201) as follows:—

343a. FALCO COMMUNIS Gm.,
var. PEALEI Ridg.
Peale's Duck Hawk.

344a. FALCO COLUMBARIUS L.,
var. SUCKLEYI Ridg.
Suckley's Pigeon Hawk.

No. 351c. The authority is HOOPES, Pr. Phila. Acad., 1873, 238, pl. 5 (Iowa).

No. 381. A dark form of *Tetrao obscurus*, from Sitka, has lately been characterized (Bull. Ess. Inst., v, 199): it is more like true *obscurus* than like var. *richardsoni*, having the broad terminal slate bar of the tail.

381b. TETRAO OBSCURUS Say,
var. FULIGINOSA (Ridg.).
Sitkan Dusky Grouse.

No. 384. A pale form of *Cupidonia*, from Texas, has lately been characterized (Bull. Ess. Inst., v, 199):—

384a. CUPIDONIA CUPIDO (L.) Bd.,
var. PALLIDICINCTA Ridg.
Texas Prairie Hen.

No. 400. A variety of this species is described by Mr. Ridgway, (Am. Nat., viii, 1874, 109), as *A. melodus* var. *circumcinctus*, having the black pectoral band complete.

400a. ÆGIALITIS MELODUS (Ord) Cab.,
var. CIRCUMCINCTUS Ridg.
Missouri Piping Plover.

No. 400bis. A new species of *Ægialitis* is described from San Francisco by Mr. Ridgway (Am. Nat., viii, 1874, 109).

400bis. ÆGIALITIS MICRORHYNCHUS Ridg.
Slender-billed Plover.

No. 401. The American form of *Æ. cantianus* may be considered

varietally distinct from the European (see Ridgway, Am. Nat., viii, 1874, 109).

401. ÆGIALITIS CANTIANUS (Lath.),
var. nivosus (Cass.) Ridgw.
Snowy Plover.

No. 402*. As intimated in the text, the North American mountain plover is perfectly distinct from the Asiatic. I have only lately seen it in breeding dress: it has *no* black pectoral band, but a transverse black coronal belt and black loral stripe.

402. EUDROMIAS MONTANUS (Towns.) Harting.
Mountain Plover.

No. 415a. It is not necessary to recognize this even by varietal name; "*M. scolopaceus*" being merely longer-billed specimens of *M. griseus*, such as may be shot out of almost any flock of the latter. The range of variation in length of bill is no greater than that occurring in *Ereunetes pusillus*, as noted in the Key, p. 254.

No. 442bis. A well known Pacific curlew, before overlooked in the Smithsonian collection, was taken at Fort Renai, Alaska, May 18, 1869, by F. Bischoff, and should be added to the last as a straggler.

[442bis.] NUMENIUS FEMORALIS Peale.
Bristle-bellied Curlew.

No. 445. According to Mr. Ridgway (Am. Nat., viii, 1874, 110), the ordinary North American glossy ibis is absolutely identical with that of the Old World. He, however, finds two other species in western United States, *I. guarauna* and *I. thalassinus*. According to this determination our species would stand as follows:—

445. IBIS FALCINELLUS Auct.
Glossy Ibis.

445bis. IBIS GUARAUNA (Linn.) Ridg.
White-faced Ibis.

445ter. IBIS THALASSINUS Ridg.
Green Ibis.

No. 448bis. I am informed by Prof. Baird that the Jabiru, of Central America, was taken some years since at Austin, Texas. It should enter the list as a straggler.

[448bis.] MYCTERIA AMERICANA L.
Jabiru.

No. 450? For occasion of the query see Key, p. 267.
No. 466. A pale form is described from California. See **Ridgway** (Am. Nat., viii, 1874, 111).

466a. RALLUS ELEGANS Aud.,
var. OBSOLETUS Ridg.
Californian Rail.

No. 470. A peculiar variety of the black rail is described from the Farallones by Mr. Ridgway (Amer. Nat., viii, 1874, 111).

470a. PORZANA JAMAICENSIS (Gm.) Cass.,
var. COTURNICULUS Bd.
Pacific Black Rail.

No. 472. The relationships of this form to the European *G. chloropus* require further investigation. It will probably stand as

472. GALLINULA CHLOROPUS Lath.,
var. GALEATA (Licht.) Hartl.

No. 479? Compare Key, p. 282.
No. 489. A lighter colored form of dusky duck, with a buff throat, is described from Florida (by Mr. Ridgway, *l. c.*), where it is resident.

489a. ANAS OBSCURA Gm.,
var. FULVIGULA Ridg.
Floridan Dusky Duck.

No. 493? The query indicates probability of only varietal distinction from No. 492.
No. 501? The query indicates probability of only varietal distinction from No. 500.
No. 575bis. A specimen of the European white-winged tern, *Hydrochelidon leucoptera*, was taken in Wisconsin, July 5, 1873, by Th. Kumlein, and presented to the Smithsonian by Dr. Brewer. This is the first instance known of its occurrence in this country.

[575bis.] HYDROCHELIDON LEUCOPTERA (Meis.) Boie.
White-winged Black Tern.

KEY TO
NORTH AMERICAN BIRDS
BY ELLIOTT COUES, M.D.

This work consists of 3G9 *Imperial 8vo pages*, and is illustrated by 6 *Steel Plates* and 238 *Woodcuts*. It is designed as a Manual or Text Book of the

BIRDS OF NORTH AMERICA,

and is an exponent of the latest views in Ornithology.

The INTRODUCTORY part gives a general account of the

ANATOMY AND CLASSIFICATION OF BIRDS

and full Explanations of all the Terms Used in ORNITHOLOGY: a *KEY to the Genera and Subgenera* follows in the form of a continuous artificial table, while a

Synopsis of Living and Fossil Birds

contains concise descriptions of every *North American Species* known at this time, with characters of the higher groups and remarks relating to forms not found in North America.

Price $7 a Copy in Full Cloth Binding.

PUBLISHED BY THE

NATURALISTS' AGENCY, Salem, Mass.

(*Prospectus furnished on application.*)

WE GIVE BELOW A FEW NOTICES OF THE PRESS WHICH WILL GIVE A GENERAL IDEA OF THE VALUE OF THIS WORK.

"The reputation of the author, who is so well known by his works on sea-birds, and for the anatomy of the loon, cannot but be increased by this production, which illustrates on every page the extent of his general information and the soundness of his judgment. The subject is treated in a manner rather different from that usually adopted by systematic writers; * * * there is a freshness and boldness in the manner in which facts are handled, which will be extremely acceptable."— *Nature* (London), May 8, 1873, p. 22.

"Mr. Coues' 'Key to North American Birds,' somewhat curiously entitled, is a very large and handsome volume, beautifully printed and profusely illustrated." — *Saturday Review* (London), Feb. 22, 1873.

"A critical examination of Dr. Coues' book reveals, it is true, here and there, slight faults of execution, but they in no way detract essentially from its value as a reliable hand-book, and one well suited to meet the wants of beginners in ornithology, while it affords at the same time a standard and convenient work of reference for advanced students and even specialists. * * * The reader is made acquainted, in a general way, with the exotic as well as North American families of the avian class. In the descriptions of the species, Dr. Coues has shown a happy skill in seizing upon such distinctions as are alone significant. * * * The geographical distribution of each species is generally fully indicated, and occasionally are added terse characterizations of their habits. * * * One of the most important features of this portion of the work, and one almost for the first time introduced into a general work on ornithology, is the critical discrimination made between species, and varieties or geographical races. Recent advances in the science have made these discriminations indispensable, and throughout the work they are rigidly and judiciously introduced. * * * The volume closes with a synopsis of all the fossil birds as yet discovered in North America, * * * being the only general exposition of this department of American ornithology that has yet been made."—*Atlantic Monthly.*

"Its author, Dr. Elliott Coues, is one of the most prominent writers on ornithology in this country, and in this volume gives the public a thorough and reliable treatise comprehending the entire subject, and so prepared that while it interests the tyro it also supplies the wants of the most advanced scientist."—*American Sportsman*, Nov. 22, 1873.

"This book will be welcomed both by the amateur and the professional ornithologist as a valuable contribution to the list of books treating of North American Birds. While Dr. Coues has modestly called his work a 'Key,' it is in reality much more than the title indicates. * * * The book will doubtless serve as a manual to many who find their recreation in observing and studying the habits of birds, and have not the means to purchase more costly works. * * * The typographical execution of the work is in every way worthy of it, and the cuts are very clear and instructive."—*The Nation*, April, 1873.

"It is what it purports to be, an exponent of the present state of American ornithology, and a perfectly reliable guide to the study of our birds. It is especially adapted to the use of students and amateurs, and is, in fact, the only *text book* or manual of ornithology arranged with a view to educational purposes. The author's high reputation as a naturalist, and his well known devotion to the department in question, guarantee the thoroughly scientific character of the work. While ranking with the best standard authorities in accuracy and completeness, this treatise presents the science of ornithology in the most attractive form, its object being to *teach*, to clear away the difficulties and explain the technicalities of the science. * * * Profusely and beautifully illustrated by the author's own hand."—*Portsmouth Journal*, April 26, 1873.

"In the present work we have the crowning result of the study of North American Birds, upon which Dr. Coues has been engaged during many years past; for, although still a young man, he has long been known as one of the most industrious of American naturalists. * * * Appears to represent his latest views upon matters of synonymy, of zoological relationship, of geographical distribution, etc. * * * Perhaps the most original feature in the book is the artificial Key to the genera of North American Birds. * * * We have gone more into detail in our notice of this work than is our custom, from our impression of its scientific and practical value, and we can cheerfully recommend it to those who wish a reliable manual of the birds of North America, in a sufficiently portable form for ready reference."—*The Independent*, March 13, 1873.

"This work, to which the author has brought the accumulated experience of years of ornithological study and the advantages of a thoroughly cultivated mind, is what it claims to be. * * * With the help of this 'Key' the veriest tyro can, with very little trouble, identify his specimens, and obtain a knowledge and understanding of American birds impossible to be found in any other work. * * * Such a book has been long wanted, and, as it has been practically tried and found of great service, it is earnestly recommended to others."—*Army and Navy Journal*, March 15, 1873.

"Dr. Elliott Coues, one of the most distinguished of our younger naturalists, has written a work on ornithology, giving a complete account of the present state and results of that science."— *The Literary Bulletin*, Oct., 1872.

"The forthcoming work of Dr. Elliott Coues on ornithology will belong distinctively to the useful class of manuals, and be especially adapted to the requirements of students, amateurs and teachers. No work of this character, professing to *teach* ornithology to the uninitiated, and susceptible of use as a text-book in educational institutions, has hitherto appeared. Those with a taste for this study, who have been deterred from its pursuit by the difficulty of mastering the technicalities in the absence of a suitable guide, will find the way made perfectly clear to them."— *The Golden Age*, Sept. 7, 1872.

"Dr. Coues has written an admirable book on North American Birds."— *Baltimore Bulletin*, March 8, 1873.

"No expense has been spared in the preparation of this volume. The woodcuts are so well executed they would easily pass for something better. The index is complete; so is the glossary."— *Chicago Times*.

"The descriptions are exceedingly complete and minute; the large number of illustrations serve to make the text more clearly understood, and the volume is a very valuable contribution to ornithology."— *Boston Journal*, Jan. 28, 1873.

"The book has been carefully prepared and contains a vast amount of information. * * * It is a book of inestimable value to the naturalist, and should be found in the library of every such person throughout the land."— *Boston Traveller*.

"A more elegant scientific publication than the 'Key to North American Birds,' just issued by this house, is not to be found. This work, of which Elliott Coues, M. D., is the author, forms a very valuable and exhaustive treatise upon the birds of the continent north of Mexico. The large number of plates and of woodcuts, renders it especially interesting, and the style of its publication is almost sumptuous."— *Boston Post*.

"Some of our distinguished men of science seem to have placed their collections and their suggestions at the service of Mr. Coues, but he is fundamentally an original explorer. Nobody can look over the beautiful book without feeling that the author has added to ornithology as well as furnished its North American 'Key.' We wish we knew enough about the subject to convict him of a few mistakes. Having, however, great respect for specialists, we never venture to intrude an opinion we have not earned the right to give by special study. It is a modest abdication of an insolent tyranny, but we make it with satisfaction. It would be cruel, perhaps, to disturb the useful superstition that notices of books are omniscient and infallible. Still, we reluctantly confess that Mr. Coues is ahead of us in his particular branch of knowledge, and we have submitted to the intolerable ignominy of learning something from him with a keen sense of pleasure. Indeed this "Key to North American Birds" is a volume which will attract all naturalists for its accuracy of description and its contributions to the work of intelligent classification."— *Boston Globe*.

"And the high commendation it has received, from competent authorities in this country and England, is even more than justified by the accuracy of description, the fulness of detail, the convenience of classification, and the admirable arrangement of the volume. * * * Dr. Coues, still a young man, and connected with the United States Army, has spent a long time in obtaining the materials for his work; but in it industry is subordinate to tact, and art to genius. He is a born naturalist. He is a close and fine observer of all natural phenomena, and were he wrecked on a rock in mid ocean he would commence scientific researches before his clothes were dry. * * * And he writes as well as he observes, in a clear, accurate style, colorless in itself, but transmitting the native hues of the objects he describes. And these qualities appear in the work before us, to enhance its value. * * * We have no hand-book of similar character, and none that occupies the place it completely fills."— *Golden Age*, July 5, 1873.

[*continued.*

happy skill in seizing upon such distinctions as are alone significant. * * * The geographical distribution of each species is generally fully indicated, and occasionally are added terse characterizations of their habits. * * * One of the most important features of this portion of the work, and one almost for the first time introduced into a general work on ornithology is the critical discrimination made between species, and varieties or geographical races. Recent advances in the science have made these discriminations indispensable, and throughout the work they are rigidly and judiciously introduced. * * * The volume closes with a synopsis of all the fossil birds as yet discovered in North America, * * * being the only general exposition of this department of American ornithology that has yet been made."—*Atlantic Monthly.*

"Its author, Dr. Elliott Coues, is one of the most prominent writers on ornithology in this country, and in this volume gives the public a thorough and reliable treatise comprehending the entire subject, and so prepared that while it interests the tyro it also supplies the wants of the most advanced scientist."—*American Sportsman*, Nov. 22, 1873.

"This book will be welcomed both by the amateur and the professional ornithologist as a valuable contribution to the list of books treating of North American Birds. While Dr. Coues has modestly called his work a 'Key,' it is in reality much more than the title indicates. * * * The book will doubtless serve as a manual to many who find their recreation in observing and studying the habits of birds, and have not the means to purchase more costly works. * * * The typographical execution of the work is in every way worthy of it, and the cuts are very clear and instructive."—*The Nation*, April, 1873.

"It is what it purports to be, an exponent of the present state of American ornithology, and a perfectly reliable guide to the study of our birds. It is especially adapted to the use of students and amateurs, and is, in fact, the only *text book* or manual of ornithology arranged with a view to educational purposes. The author's high reputation as a naturalist, and his well known devotion to the department in question, guarantee the thoroughly scientific character of the work. While ranking with the best standard authorities in accuracy and completeness, this treatise presents the science of ornithology in the most attractive form, its object being to *teach*, to clear away the difficulties and explain the technicalities of the science. * * * Profusely and beautifully illustrated by the author's own hand."—*Portsmouth Journal*, April 26, 1873.

"In the present work we have the crowning result of the study of North American Birds, upon which Dr. Coues has been engaged during many years past; for, although still a young man, he has long been known as one of the most industrious of American naturalists. * * * Appears to represent his latest views upon matters of synonymy, of zoological relationship, of geographical distribution, etc. * * * Perhaps the most original feature in the book is the artificial Key to the genera of North American Birds. * * * We have gone more into detail in our notice of this work than is our custom, from our impression of its scientific and practical value, and we can cheerfully recommend it to those who wish a reliable manual of the birds of North America, in a sufficiently portable form for ready reference."—*The Independent*, March 13, 1873.

"This work, to which the author has brought the accumulated experience of years of ornithological study and the advantages of a thoroughly cultivated mind, is what it claims to be. * * * With the help of this 'Key' the veriest tyro can, with very little trouble, identify his specimens, and obtain a knowledge and understanding of American birds impossible to be found in any other work. * * * Such a book has been long wanted, and, as it has been practically tried and found of great service, it is earnestly recommended to others."—*Army and Navy Journal*, March 15, 1873.

"Dr. Elliott Coues, one of the most distinguished of our younger naturalists,

[*continued.*

[*continued*.

has written a work on ornithology, giving a complete account of the present state and results of that science."— *The Literary Bulletin*, Oct., 1872.

"The forthcoming work of Dr. Elliott Coues on ornithology will belong distinctively to the useful class of manuals, and be especially adapted to the requirements of students, amateurs and teachers. No work of this character, professing to *teach* ornithology to the uninitiated, and susceptible of use as a text-book in educational institutions, has hitherto appeared. Those with a taste for this study, who have been deterred from its pursuit by the difficulty of mastering the technicalities in the absence of a suitable guide, will find the way made perfectly clear to them."— *The Golden Age*, Sept. 7, 1872.

"Dr. Coues has written an admirable book on North American Birds."— *Baltimore Bulletin*, March 8, 1873.

"No expense has been spared in the preparation of this volume. The woodcuts are so well executed they would easily pass for something better. The index is complete; so is the glossary."— *Chicago Times*.

"The descriptions are exceedingly complete and minute; the large number of illustrations serve to make the text more clearly understood, and the volume is a very valuable contribution to ornithology."— *Boston Journal*, Jan. 28, 1873.

"The book has been carefully prepared and contains a vast amount of information. * * * It is a book of inestimable value to the naturalist, and should be found in the library of every such person throughout the land."— *Boston Traveller*.

"A more elegant scientific publication than the 'Key to North American Birds,' just issued by this house, is not to be found. This work, of which Elliott Coues, M. D., is the author, forms a very valuable and exhaustive treatise upon the birds of the continent north of Mexico. The large number of plates and of woodcuts, renders it especially interesting, and the style of its publication is almost sumptuous."— *Boston Post*.

"Some of our distinguished men of science seem to have placed their collections and their suggestions at the service of Mr. Coues, but he is fundamentally an original explorer. Nobody can look over the beautiful book without feeling that the author has added to ornithology as well as furnished its North American 'Key.' We wish we knew enough about the subject to convict him of a few mistakes. Having, however, great respect for specialists, we never venture to intrude an opinion we have not earned the right to give by special study. It is a modest abdication of an insolent tyranny, but we make it with satisfaction. It would be cruel, perhaps, to disturb the useful superstition that notices of books are omniscient and infallible. Still, we reluctantly confess that Mr. Coues is ahead of us in his particular branch of knowledge, and we have submitted to the intolerable ignominy of learning something from him with a keen sense of pleasure. Indeed this "Key to North American Birds" is a volume which will attract all naturalists for its accuracy of description and its contributions to the work of intelligent classification."— *Boston Globe*.

"And the high commendation it has received, from competent authorities in this country and England, is even more than justified by the accuracy of description, the fulness of detail, the convenience of classification, and the admirable arrangement of the volume. * * * Dr. Coues, still a young man, and connected with the United States Army, has spent a long time in obtaining the materials for his work; but in it industry is subordinate to tact, and art to genius. He is a born naturalist. He is a close and fine observer of all natural phenomena, and were he wrecked on a rock in mid ocean he would commence scientific researches before his clothes were dry. * * * And he writes as well as he observes, in a clear, accurate style, colorless in itself, but transmitting the native hues of the objects he describes. And these qualities appear in the work before us, to enhance its value. * * * We have no hand-book of similar character, and none that occupies the place it completely fills."— *Golden Age*, July 5, 1873.

www.ingramcontent.com/pod-product-compliance
Lightning Source LLC
Chambersburg PA
CBHW030351170426
43202CB00010B/1342